CALIFORNIA NATURAL HISTORY GUIDES

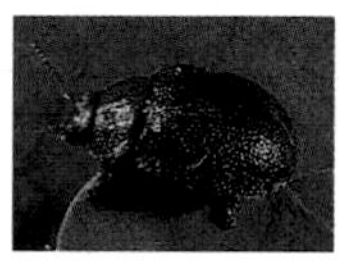

INTRODUCTION TO CALIFORNIA BEETLES

California Natural History Guides

Phyllis M. Faber and Bruce M. Pavlik, General Editors

Introduction to CALIFORNIA BEETLES

Arthur V. Evans
James N. Hogue

UNIV... PRESS
Berkeley Los Angeles London

We dedicate this book to our wives, Paula and Cheryl, for their support of our entomological pursuits.

California Natural History Guides No. 78

University of California Press
Berkeley and Los Angeles, California

University of California Press, Ltd.
London, England

Library of Congress Cataloging-in-Publication Data

Evans, Arthur V.
Introduction to California beetles / Arthur V. Evans and James N. Hogue.
p. cm. — (California natural history guides ; 78)
Includes bibliographical references and index.
ISBN 0–520-24034-0 (case) — ISBN 0–520-24035-9 (paper)
1. Beetles — California — Identification. I. Hogue, James N., 1961– II. Title. III. Series.

QL584.C2E94 2004
595.76′09794—dc22 2003061782

Manufactured in China
10 09 08 07 06 05 04
10 9 8 7 6 5 4 3 2 1

The paper used in this publication meets the minimum requirements of ANSI/NISO Z39.48–1992 (R 1997) (*Permanence of Paper*). ♾

Cover: A female rain beetle *(Pleocoma badia badia)*. Photography by J. N. Hogue.

The publisher gratefully acknowledges the generous contributions to this book provided by

the Gordon and Betty Moore Fund
in Environmental Studies

and

the Heller Charitable and Educational Fund

CONTENTS

Acknowledgments ix
Introduction xv

1. A BRIEF HISTORY OF BEETLE STUDY IN CALIFORNIA 1

2. FORM, DIVERSITY, AND CLASSIFICATION 25

3. THE LIVES OF BEETLES 61

4. DISTRIBUTION OF CALIFORNIA BEETLES 83

5. BEETLES OF SPECIAL INTEREST 97

6. COMMON AND CONSPICUOUS FAMILIES OF CALIFORNIA BEETLES 137

Ground Beetles and Tiger Beetles (Carabidae) 138
Whirligig Beetles (Gyrinidae) 141
Predaceous Diving Beetles (Dytiscidae) 144
Water Scavenger Beetles (Hydrophilidae) 147
Clown Beetles (Histeridae) 151
Carrion Beetles (Silphidae) 154
Rove Beetles (Staphylinidae) 156
Rain Beetles (Pleocomidae) 158

Scarab Beetles, Dung Beetles, May Beetles, June Beetles, and Chafers (Scarabaeidae) 162
Jewel Beetles, or Metallic Wood-boring Beetles (Buprestidae) 166
Click Beetles (Elateridae) 169
Phengodid Beetles (Phengodidae) 172
Fireflies and Glowworms (Lampyridae) 174
Soldier Beetles (Cantharidae) 176
Skin Beetles (Dermestidae) 179
Powder-post Beetles, Twig Borers, or Branch Borers (Bostrichidae) 181
Checkered Beetles (Cleridae) 184
Lady Beetles (Coccinellidae) 186
Darkling Beetles (Tenebrionidae) 189
Blister Beetles (Meloidae) 192
Longhorn Beetles (Cerambycidae) 194
Leaf Beetles (Chrysomelidae) 200
Weevils or Snout Beetles, Ambrosia Beetles, and Bark Beetles (Curculionidae) 204

7. STUDYING BEETLES **209**

Checklist of North American Beetle Families 255
California's Sensitive Beetles 261
Collections, Societies, and Other Resources 265
Selected References 268
Index 279

ACKNOWLEDGMENTS

Any book of this nature must necessarily be a collaborative effort, requiring the cooperation of numerous individuals working in public and private institutions. We take this opportunity to thank those friends and colleagues who so generously offered their energy and expertise in support of this project.

We thank the following for supplying copies of literature that contributed substantially to the distribution, taxonomic, and biographical sections of the book: Nancy Baker and Dr. Steve Lingafelter (National Museum of Natural History, Smithsonian Institution, Washington, D.C.); Dr. Brian Brown and Victoria Brown (Natural History Museum of Los Angeles County); Dr. John Chemsak (University of California at Berkeley); Dave Faulkner (San Diego); Dr. Robert Gordon (Willow City, North Dakota); Dr. Lee Herman (American Museum of Natural History, New York); Dr. Al Newton and Dr. Margaret Thayer (Field Museum of Natural History, Chicago, Illinois); Dr. John Pinto (University of California at Riverside); Dr. Rick Westcott (Salem, Oregon); and Dr. Dan Young (University of Wisconsin at Madison).

The following individuals kindly provided timely specimen records of California beetles from collections in their care: Cheryl Barr (University of California at Berkeley); Roberta Brett (California Academy of Sciences, San Francisco); Brian Harris (Natural History Museum of Los Angeles County); and Dr. Steve Heydon (University of California at Davis).

We are grateful to the librarians and entomologists at the National Museum of Natural History, Smithsonian Institution, Washington, D.C., who assisted Arthur V. Evans in tracking down pertinent literature and numerous other kindnesses. Dr. Steve Lingafelter and Dr. Warren Steiner identified troublesome California specimens. We are especially indebted to Dr. David Furth for making the resources of the museum available to one of us, Arthur V. Evans, by sponsoring his appointment as a research associate. Similarly, we greatly appreciate the many invaluable resources made available to James N. Hogue by the Department of Biology, California State University at Northridge, and the Entomology Section of the Natural History Museum of Los Angeles County.

Dr. Paul Skelley (Florida State Collection of Arthropods, Gainesville) saved us tremendous amounts of time searching the literature by providing PDF files of the then unpublished second volume of *American Beetles.* The following contributing authors graciously provided early drafts of their manuscripts: Dr. Rolf Aalbu (Sacramento, California); Dr. Mary Liz Jameson and Dr. Brett Ratcliffe (University of Nebraska State Museum, Lincoln); and Dr. Keith Phillips (Western Kentucky University, Bowling Green, Kentucky).

The following people cordially supplied unpublished state checklists generated from their personal data bases of North American species records: Dr. Larry Bezark and Dr. Richard Penrose (California Department of Food and Agriculture, Sacramento) (Cerambycidae); Ed Riley (Texas A & M University, College Station, Texas) (Chrysomelidae); Dr. Michael Caterino (Santa Barbara Museum of Natural History, Santa Barbara, California) (Curculionidae); Dr. Robert Rabaglia (Maryland Department of Agriculture, Baltimore) (Scolytinae).

Dr. Yves Alarie and Jennifer Babin (Laurentian University, Sudbury, Ontario, Canada) clarified taxonomic questions regarding Gyrinidae and verified species records for the state. Dr. Rolf Aalbu and Jacques Rifkind (North Hollywood, California) reviewed and updated our species lists for Tenebrionidae, Zopheridae, and Cleridae.

Dr. Michael Ivie (Montana State University, Bozeman) deserves special credit for bringing to our attention the seminal contribution of nineteenth-century Russian entomologists to the history of California coleopterology. He provided key historical and biographical references that inspired the development of the chapter outlining the history of the study of beetles in California.

Chapter 5, "Beetles of Special Interest," was greatly enhanced by Dr. Robert Dowell (California Department of Food and Agriculture, Sacramento), who provided valuable information on local and exotic pest species. Dr. Rosser Garrison and James Wiseman (Los Angeles County Agricultural Commissioner and Weights and Measures) reviewed an early draft of the chapter and provided detailed documentation regarding eucalyptus pests and Japanese beetles in California. Conversations and correspondence with Dr. Laura Merrill (San Bernardino National Forest) helped to focus the section on California forest pests.

Andrew Smith (University of Nebraska State Museum, Lincoln) generously reviewed the essays covering the scarabaeoid families (Pleocomidae and Scarabaeidae).

Unless otherwise noted, all of the color photographs are our own. All of the line drawings were executed by James N. Hogue. We respectfully acknowledge the University of California Press for permission to reproduce the map depicting the landforms and natural areas of California. Permissions to use black-and-white photographs of coleopterologists were provided by the following: Pacific Coast Entomological Society; Alaska and Polar Regions Department, Elmer E. Rasmuson Library, University of Alaska at Fairbanks; Department of Entomology, National Museum of Natural History, Smithsonian Institution; and Dr. John Chemsak, Curator Emeritus, Essig Museum of Entomology, University of California at Berkeley.

We especially thank Doris Kretschmer, Senior Editor at the University of California Press. Her early encouragement and enthusiasm for a popular introduction to California beetles were invaluable during the earliest stages of the book's development. Her comments regarding earlier drafts of the work were instrumental in helping us to find our voices for sharing our passion for beetles in this book.

We are particularly grateful to the individuals assigned by the University of California Press to read the penultimate draft of the book. Dr. Bruce Pavlik (Mills College, Oakland, California) made numerous cogent observations that helped to maintain strict scientific accuracy of the book. The diligent and careful reading of Dr. David Wood, Professor Emeritus (University of California at Berkeley), resulted in numerous suggestions and corrections that increased its overall clarity. Finally, Dr. Michael Caterino (Santa Barbara Museum of Natural History) thoroughly and thoughtfully reviewed the draft, adding substantially to the accuracy of the biological and taxonomic aspects of the book. His enthusiastic and scholarly approach to the study of California's beetles (see the section "Societies and Web Sites Promoting the Study of Beetles") remains an inspiration to us.

Dr. Charles Bellamy (Department of Food and Agriculture, Sacramento) responded to numerous questions regarding California Buprestidae, as well as several other families, and tracked down many references and locality records. His friendship and unflagging support over the years, combined with his keen taxonomic acumen and wry humor, are always appreciated and greatly admired.

Paula Evans (Richmond, Virginia) eagerly read every draft of the manuscript and made numerous corrections and suggestions that added tremendously to the overall tone, readability, and organization of the book. Without her love and curiosity of nature, enthusiasm for insects, and unstinting support and encouragement, this book would not have been possible.

We share the success of *Introduction to California Beetles* with all the aforementioned individuals, but the responsibility for any and all of its shortcomings, misrepresentations, and inaccuracies is entirely our own.

Arthur V. Evans
James N. Hogue
January 2003

INTRODUCTION

We live in the Age of Beetles. There are more beetles than any other group of organisms, with perhaps 350,000 different species known worldwide. Eight thousand or more of those live in California, many of which are found nowhere else on the planet. Unlike flowering plants, birds, and butterflies, we still do not have a complete listing of all the beetles in the state for two main reasons: the sheer number of species and the vast regions still unexplored, awaiting surveys by professional and amateur beetle specialists.

Introduction to California Beetles is the first book to focus on the state's beetle fauna. This book provides an overview of the natural history, ecology, body form, classification, and study of the state's beetles. It was written with biologists, naturalists, educators, students, gardeners, and other outdoor enthusiasts in mind. Residents and visitors alike in the Golden State frequently encounter large, colorful, or conspicuous beetles and marvel at their engaging, sometimes mysterious behaviors. However, until the publication of this book, there were few alternatives for obtaining information about these captivating creatures because few popular regional books on beetles exist anywhere. Natural history information and identification guides for most beetles are found in widely scattered articles published in scientific journals. This information is generally difficult for nonspecialists to obtain and even harder to understand since it is written by specialists for their colleagues. This book attempts to fill that void by distilling the scientific literature and presenting it in an introductory format. The information is accessible to a broad audience, yet authoritative so that it can also serve as a useful reference for the specialist.

The heart of the book lies in the essays and descriptions of some of the larger, more conspicuous groups of the state's beetles. Each family presented is illustrated with a razor-sharp color photo of a living beetle. We present details as to where and how these fascinating animals live and suggestions for their observation and study. One of the most appealing aspects of studying California beetles is the fact that they can be found virtually anywhere, anytime. Whether exploring your own backyard, visiting a state park, or hiking in a national forest, every outing throughout the year will reveal something new and exciting about beetles and their role in the natural world. And, because beetles are the perfect ambassadors for environmental awareness, we present a section on their care and maintenance in captivity for use in educational programming.

The book also includes a brief and colorful history of beetle study in the state, beginning with the native Americans and followed by the Spanish mission builders and the Russian explorers. American military expansion in the West fueled tremendous growth in the study of California beetles. Later, as the state flexed its agricultural might, new methods of pest control were required, and California became the birthplace of modern biological control. Talented and enthusiastic collectors flocked to the state, most of whom were self-taught entomologists. Their work formed the foundations for the collections of today's important museums and universities, inspiring and training present and future generations of California's beetle workers.

Another chapter focuses on beetles of special interest: fossil species that provide a glimpse of ancient habitats, ravenous home and pantry pests, destructive forest pests, endangered species, and more. Readers who want to learn more about California beetles are urged to consult the selected references for further information on their identification and study. The appendices include the latest checklist of families in North America, all but 13 occurring in California, and a catalog of the state's sensitive, threatened, and endangered species. These are followed by a listing of important and publicly held beetle collections and a list of societies and their Web sites promoting the study of beetles.

California is not only a melting pot for people, but also one for beetles. Although the state is isolated by desert, mountain, and sea, the ancient origins of California's beetle fauna are rooted in both the temperate Old World and New World tropics. And with its variety of climates and steady influx of people and goods from around the world, the potential for successful introductions of exotic beetles—some of which might achieve pest status—remains high.

The authors have more than 50 years of combined experience observing, collecting, photographing, and marveling at beetles, and this book is the culmination of those energies. We hope that *Introduction to California Beetles* will inspire others to delight in discovering these and other incredible animals with which we share our planet. Join us as we embark into the fascinating world of California beetles. It is the beginning of a lifetime journey of wonder and awe.

CHAPTER 1
A BRIEF HISTORY OF BEETLE STUDY IN CALIFORNIA

Beetles have long captured the imagination of California's human inhabitants. No doubt the earliest interest in California beetles came from the indigenous peoples who lived in the region for over 10,000 years, long before the arrival of the Europeans. Beetles played an important and practical role in the lives of these indigenous peoples, possibly as medicine and most certainly as food. Later, Spanish mission builders introduced agriculture to the state but had little direct interest in California beetles. The activities of the Spanish, however, did result in the introduction of several kinds of important beetle pests to California from the Old World.

The first scientific studies of California beetles began with the Russian occupation of northern California in the early 1800s. The influx of Americans and Europeans to California in the wake of the Gold Rush in the mid-1800s led to an increased interest in California's unique beetles. With the rapid increase of the state's agricultural prowess and subsequent pest infestations, the stage was set for the establishment of research institutions, universities, and societies fostering the study of California beetles. The early beetle collections of these institutions were built largely by dedicated amateurs who conducted extensive field work throughout the state on their own time and money.

Native Americans and Beetles

California's Native Americans considered the larvae of the Pine Sawyer *(Ergates spiculatus)* and the California Prionus *(Prionus californicus)* to be delicacies. These sausage-sized grubs are an excellent source of fat and were removed from logs or stumps to be cooked or eaten raw. Other longhorn beetle (Cerambycidae) larvae, including the Ribbed Pine Borer *(Rhagium inquisitor),* Nautical Borer *(Xylotrechus nauticus),* Spotted Pine Sawyer *(Monochamus maculosus),* and

Black Pine Sawyer *(M. scutellatus)*, were also consumed. Weevil grubs infesting stores of acorn and pine nuts not only provided an additional source of protein, but they probably enhanced the nutty and oily flavor of the meal. Adult Striped June Beetles (*Polyphylla crinita,* Scarabaeidae) and other common scarabs attracted to evening campfires were frequently consumed.

Not all Native American encounters with beetles were of the culinary sort. Large black and bumbling darkling beetles (Tenebrionidae) were certainly known to the Mojave by their offensive smell and were dubbed *humahnana.* The Juaneño people referred to bright red ladybugs (Coccinellidae) as *coronnes,* and yellow species were called *tepis.* Interestingly, these were the same words used to refer to the first and second wives of their chief.

European Colonization

The Spanish settlements in what is now California began with the establishment of Mission San Diego in 1769 and stretched northward to Sonoma. The influence of the mission system began to deteriorate in 1834 and abruptly ended in 1846. The primary aim of each mission was the development of agriculture to provide food for the native people and other mission workers. To this end the mission fathers brought with them all kinds of plant materials from Europe and elsewhere, including seeds, vines, and rootstock. Hidden among these materials were some of the first beetles introduced to western North America. European grain pests such as the Granary Weevil *(Sitophilus granarius)* and the Rice Weevil (*Sitophilus oryzae,* Curculionidae) were found sealed in the adobe bricks of the Santo Domingo mission in Baja California, built in 1775, which suggests that these and other pests were introduced to California with the establishment of the San Diego mission.

Beginning as early as 1779, whaling vessels and hide ships carrying furs and tallow from Asia and Europe landed at the harbors of San Francisco and Monterey to obtain supplies. From these ships were probably the first introductions of European ham beetles (*Necrobia,* Cleridae), skin beetles (*Anthrenus* and *Dermestes,* Dermestidae), and spider beetles (*Ptinus,* Anobiidae). Native skin and hide beetles were also spread up and down the coast by these ships and were very likely introduced to other ports elsewhere in the world.

Early Russian Influences

Unlike the Spanish, whose interests in California were primarily mission building, the Russians were very much interested in the exploration of the region's natural history. Russian America stretched from Alaska southward to coastal northern California. Fort Ross (pl. 1), located north of San Fran-

Plate 1. Fort Ross, located north of San Francisco, was a noted trading center and farming community established by the Russians in 1812 and was the first center of California beetle study.

Figure 1. Physician and naturalist Johann Freldrich Eschscholtz (1793 to 1831) was the first person to formally collect and describe beetles in California. Portrait from Eschscholtz's *Zoologischer Atlas,* Rare B0227, Alaska and Polar Regions Archives, Rasmuson Library, University of Alaska at Fairbanks.

cisco, was a noted trading center and farming community established by the Russians in 1812 in what was then called New California. The community became a focal point for many Russian insect collectors who scoured the territory between Bodega Bay and Mount St. Helena for specimens, especially beetles. Fort Ross was abandoned in 1841, but not before several prominent entomologists and naturalists had studied beetles from the region.

Johann Freidrich Eschscholtz (1793 to 1831) (fig. 1) made two voyages to California, courtesy of the Imperial Russian Navy. In 1815 Eschscholtz arrived on the brig *Rurik* as the ship's physician and naturalist. He collected for only one month in the vicinity of San Francisco and was accompanied by botanist Adelbert von Chamisso. On this trip, the first specimens of the California poppy were collected. Chamisso named the species *Eschscholtzia californica,* in honor of his friend and colleague.

Eschscholtz returned to San Francisco Bay in September 1824. His collections on this trip were made in the vicinity of San Francisco, Santa Clara, San Rafael, Bodega Bay, and the lower Sacramento River. He spent several days in the vicinity of Fort Ross, where he obtained a large series of nearly 100 species of beetles.

Eschscholtz later visited the French naturalist Pierre Francois Marie Auguste Dejean in Paris. Dejean had accumulated the greatest collection of beetles in the world, with over 22,000 species. Here, Eschscholtz penned many of the descriptions of the beetles he collected during his last trip to California. After Eschscholtz's death, Dejean published many of Eschscholtz's descriptions in 1836. Although Dejean attributed the descriptions to Eschscholtz, the rules of zoological nomenclature dictate that he and not Eschscholtz is the author of the species.

The Governor of Finland, Carl Gustov von Mannerheim (1804 to 1854), prepared and described much of the beetle material accumulated by Russian museums, especially those collected on expeditions to Siberia, Alaska, and California. He wrote two of the first papers on California beetles in 1840 and 1843, describing hundreds of species. His collectors included the governor of Russian America (F. P. Wrangell), two physicians in the Russian American Company (E. L. Blaschke and F. Fischer), the overseer at Fort Ross (G. Tschernikh), and an entomologist (I. G. Vosnesensky).

Ilya Gavrilovich Vosnesensky (1816 to 1871) was the only Russian to visit California who was trained as an entomologist. He served as an apprentice to E. Ménétriés at the Imperial Academy of Sciences at St. Petersburg. The Imperial Academy sent Vosnesensky to Russian America specifically for the purpose of collecting insects. He arrived at Fort Ross in July 1840 and remained there until September 1841, collecting in the vicinities of San Francisco, Russian River, Bodega Bay, and New Helvetia. New Helvetia, now the site of Sacramento, was founded by the Swede John Sutter, who established Fort Sutter there in 1841. Russian entomologists C.V. von Mannerheim, V. I. Motschulsky, and E. Ménétriés described Vosnesensky's beetles.

Military Outposts and Entomological Exploration in California

Logistical difficulties excluded California and much of the West from any thorough explorations by American entomologists in the early nineteenth century. Collections made by a few adventurous collectors during railroad surveys and at military garrisons during the mid-1800s, however, yielded many exciting new beetles from California. American military outposts in California served as early centers of collecting activity for eastern entomologists employed as army surgeons.

Fort Tejon was a military post built in 1852 and was established to protect immigrants from bandits and renegades. Located in the mountains between Bakersfield and Los Angeles, Fort Tejon was a popular and fertile hunting ground for beetle collectors of the day. Hungarian zoologist and botanist John Xantus de Vesey collected there in 1857 to 1858. John L. LeConte collected beetles in the region in 1850 and described many beetles collected there by Xantus. G.H. Horn also visited Fort Tejon and collected a large number of beetles there sometime between 1863 and 1867.

Horn also collected numerous beetles at another important early entomological site in California, Camp Independence. Located east of the Sierra Nevada in Inyo County where what is now the town of Independence, Camp Independence was established on Oak Creek in the Owens Valley in 1862. The camp was leveled and immediately rebuilt after the earthquake of 1872 but was later abandoned in 1877.

American entomology at this time was frustratingly difficult, requiring consultation of European literature, specimens, and entomologists. The flow of specimens outside the country fueled efforts to build and maintain large, permanent, and comprehensive insect collections staffed by professional entomologists. By the end of the Civil War and the completion of the first transcontinental railway in 1869, the stream of specimens from the West to entomologists and the scientific institutions of eastern United States increased dramatically. By the 1880s, institutions such as Harvard University, the Philadelphia Academy of Sciences, the Smithsonian Institution, and their publications established American entomology as a science comparable to that of Europe.

Centers for Beetle Study in California

Beginning with the discovery of gold in 1849, enormous numbers of people from all walks of life, including naturalists, immigrated to California. As California's population grew, so did the state's agricultural industry, particularly its fruit culture. This chain of events set into motion the establishment of California's research institutions and universities, first in the northern part of the state and then spreading southward. With its institutions approaching the status of those in the east, California continued to attract enthusiastic researchers and collectors from the rest of country.

California Academy of Sciences

Stemming the flow of western North America's insects to the institutions of the east was the establishment of the state's oldest scientific society, the California Academy of Natural Sciences. Founded in San Francisco in 1853, the Academy later changed its name to the California Academy of Sciences in 1868. Construction of its first permanent buildings to store collections began on Market Street in 1891. The Academy's collections and libraries provided the first focal point for California beetle study.

Unfortunately the Academy's collections and libraries were almost completely destroyed in the fire following the San Francisco earthquake of 1906. Temporarily headquartered in the Security Building on Market Street, the Academy immediately set out to rebuild its collections and library. The new buildings were completed in 1915 at their present location in Golden Gate Park.

Since 1933, the monthly meetings of the Pacific Coast Entomological Society have been held in the Academy's Entomology Department. Today the Academy houses the largest insect collection in California, with more than 12 million specimens, and is the site of the most important beetle collection west of the Mississippi River.

University of California

Founded at Berkeley in 1868, the University of California was established as California's foundation for agricultural investigation and teaching in the state. As the state's agricultural prowess grew, so did its insect pest problems. Special instruction in entomology at the university began in 1882, and the Department of Entomology and Parasitology was established in 1920. Agricultural Experiment Stations were set up throughout the state to study regional agricultural pest problems, especially those affecting citrus groves. The California Insect Survey was started in 1940 by the Agricultural Experiment Station in an effort to build a comprehensive collection of the state's insect fauna and facilitate projects in applied entomology in California. The Essig Museum of Entomology at the University of California at Berkeley was founded as the collection component of this survey and currently contains over five million specimens. The University of California system now has two other campuses (Davis and Riverside) that house important California beetle collections.

San Diego Museum of Natural History

About the same time the new California Academy of Sciences reopened its doors, two other important entomological institutions were founded in Los Angeles and San Diego. The San Diego Natural History Museum traces its roots to an enthusiastic group of amateur naturalists who formed the San Diego Society of Natural History in 1874. The Society opened its first exhibits in a hotel in 1912 and moved to Balboa Park five years later. Its present facility was built at that location in 1933. This small but historically important beetle collection contains species from the San Diego region and Baja California.

Natural History Museum of Los Angeles County

Under the auspices of the county of Los Angeles, the Museum of History, Science, and Art formally opened in November 1913. The early reputation of the museum was built on the internationally renowned Pleistocene deposits, including insects, recovered from the La Brea Tar Pits. With the departure of the Art Department in the early 1960s, the museum became known as the Los Angeles County Museum of Natural History and later as the present Natural History Museum of Los Angeles County.

During the 1920s and 1930s, the museum sponsored the annual Butterfly Show featuring numerous insect displays. This tradition continues today with the annual Insect Fair held in May. The museum is also the site of one of the country's largest insect zoos, attracting nearly a quarter million visitors each year.

With eight million specimens, the Entomology Section houses the second largest insect collection in California. Over the years it has absorbed the entomological collections of the University of California at Los Angeles, University of Southern California, and Stanford University. The collection is a repository for the California Channel Island survey conducted in the 1930s and 1940s. The C. D. Nagano collection of tiger beetles is deposited here. In 1995, one of the authors (Evans) added his worldwide collection of scarab beetles, totaling some 40,000 specimens, to the museum. The Lorquin Entomological Society, named after the famed French naturalist who explored California in the mid-1800s, is affiliated with the Entomology Section and has held its meetings at the museum since 1927.

California Beetle Workers

California, with its profusion of beetles and habitats, has inspired professional entomologists and dedicated amateur naturalists for nearly 200 years. After the Russian explorers left California in the 1840s, several European and American coleopterists made significant contributions to our understanding of the state's beetle fauna. Although a few were trained as entomologists, most were physicians, tradesmen, or independently wealthy naturalists building their own private beetle collections (fig. 2). These private collections eventually found their way into museums and universities throughout the country and form the foundation of our understanding of beetle classification, distribution, and biology. The lives, interests, and contributions of some of these men are briefly presented below.

John Lawrence LeConte (1825 to 1883)

LeConte (fig. 3) was one of the most eminent American entomologists of the nineteenth century and was recognized as the premier authority on North American beetles throughout the world. He was a pioneer in the classification of North American beetles, earning him the reputation as the father of American beetle study.

Figure 2. Taken in San Francisco in 1888, this photo includes some of the most prominent beetle collectors of the day. Standing, from left to right, are Albert Koebele, James H. Behrens, and Thomas Lincoln Casey. Seated, from left to right, are James Rivers, George W. Dunn, Charles Fuchs, and W.G.W. Harford. Photo from Grinnell (1914).

LeConte's mother died only a few months after his birth in New York City. His father, Major John E. LeConte, who had published several papers on butterflies and beetles, raised his son alone. The major instilled in young LeConte a passion for natural history at an early age. Publishing his first three papers on beetles in 1844 at the age of 19, LeConte would go on to describe some 5,000 species of beetles in his lifetime, many of which are represented in California.

In one of his first papers LeConte established himself as an outspoken advocate of the need for American entomologists

Figure 3. John Lawrence LeConte (1825 to 1883) was one of the most eminent American entomologists of the nineteenth century. A pioneer in the classification of North American beetles, LeConte is considered the father of American beetle study. Photo from Scudder (1884).

to describe American insects. He deplored the practice of sending specimens to Europe for determination or description and in 1845 announced "America's Entomological Declaration of Independence."

While training as a physician at the college of Physicians and Surgeons in New York, LeConte spent much of his time traveling about the United States collecting beetles. After receiving his medical degree in 1846, LeConte never entered practice, although he later served the army as a surgeon. LeConte traveled to California via the Isthmus of Panama in 1849 and collected insects from areas around San Francisco and San Diego. He once sent 10,000 beetles preserved in alcohol from San Francisco to his father in New York. Another 20,000 or so specimens were apparently lost in the fire that ravaged San Francisco in 1852.

Working with another pioneer entomologist, S. S. Haldeman, LeConte revised F. E. Melsheimer's *Catalogue of the Coleoptera of the United States.* Published in 1853 by the Smithsonian Institution, the catalog marked the first organized study of beetles in the United States and listed 540 species of beetles from California. Most importantly the catalog served as the foundation for subsequent catalogs of North American beetles.

LeConte examined some of the earliest California beetle specimens available to eastern entomologists. These specimens included those collected by Dr. Charles Pickering and Mr. Titian Peale, both ship's naturalists that visited California in 1841 as part of an exploratory expedition conducted by the U.S. Navy under the command of Captain Charles Wilkes. Other specimens were acquired from eastern private collectors such as J. Wittick and J. Child who had both visited the area around Sacramento. LeConte also examined material collected by the Russians Eschscholtz and Mannerheim, sent to him by Baron Chaudoir and Colonel Motschulsky. LeConte recognized early on that California constituted a unique zoological region based on the fact that many of its beetles were found nowhere else in nature.

LeConte's *Classification of the Coleoptera of North America* appeared in 1861 and 1862, but the Civil War curbed further publications on beetles until 1873. During the interim, LeConte served as an Army Medical Corps surgeon with the California volunteers in the infantry during the Civil War from 1862 to 1866. Once again he took this opportunity to collect many new and unusual species of beetles in the state.

In 1878, LeConte was appointed assistant director of the United States Mint in Philadelphia, a position he held until his death in 1883, the same year his final publication appeared. It was a revision of his earlier work on the classification of the Coleoptera, written with his student and longtime friend and colleague, G. H. Horn. LeConte's collection of beetles, including hundreds of California specimens, now resides in Harvard University's Museum of Comparative Zoology in Cambridge, Massachusetts.

George Henry Horn (1840 to 1897)

Born and educated in Philadelphia, George Henry Horn (fig. 4) received his medical degree from the University of Pennsylvania in 1861. His first paper on beetles appeared the same

Figure 4. George Henry Horn (1840 to 1897) arrived in California during the Civil War and collected beetles throughout the state. Upon LeConte's death in 1883, Horn became the preeminent coleopterist in North America. Photo from Skinner (1898).

year and attracted the attention of J.L. LeConte, who sought him out. The two men soon became friends and remained collaborators for more than 20 years.

Horn arrived in California during the Civil War, where Governor Leland Stanford commissioned him assistant surgeon in the Second Calvary, California Volunteers. He took his oath at Camp Independence in 1863. Like LeConte, he took the opportunity to collect beetles throughout much of California, as well as in parts of Arizona and Nevada. During his three years in the West, Horn collected at Fort Tejon, Fort Yuma (on the Colorado River, Arizona), Fort Crook (Shasta County), Surprise Valley (Modoc County), Warner's Ranch (San Diego County), and the Sacramento Valley. In 1865 he achieved the rank of major and eventually returned to Philadelphia in 1866. There he established his medical practice and became an obstetrician.

With a thriving practice, Horn's entomological pursuits were conducted in his spare time. He used several characters to describe and distinguish species and preferred to publish his findings as monographs so that all of the species in a group known to him could be identified. He published a total of 265 papers on beetles, creating 154 new genera and describing over 1,600 new species, many of which are represented in California. Upon LeConte's death in 1883, Horn became the preeminent coleopterist in North America.

Horn traveled once again to California in spring of 1893, where he met with other California coleopterists, including H.C. Fall, and was introduced at the May meeting of the California Academy of Sciences in San Francisco.

Thomas Lincoln Casey (1857 to 1925)

Thomas Lincoln Casey (fig. 5) was born into a military family in West Point, New York. He entered the United States Military Academy and graduated with honors in 1879. He received a commission in the Engineer Corps of the U.S. Army as an engineer and retired in 1912 with the rank of colonel. He was stationed in South Africa and various parts of the United States, including California from 1885 to 1886. While in California, Casey collected from San Diego to Eureka and secured a wealth of material. He preferred to work primarily on obscure families of beetles, however, and many of his species are unfamiliar to many entomologists.

Although he collected many beetles himself, Casey amassed most of his remarkable collection by purchasing specimens from other collectors. His collection and library grew so large that Casey was forced to rent two apartments in Washington, D.C., one for living in and the other for his beetles, papers, and books. His library and collection, including thousands of California beetles, are now part of the National Museum of Natural History in Washington, D.C.

Casey published numerous papers on engineering, conchology, and astronomy. His first papers on beetles appeared in 1884. During the next 40 years Casey described nearly 10,000 new species, publishing almost 9,000 pages on their taxonomy and biology.

Figure 5. Thomas Lincoln Casey (1857 to 1925) collected from San Diego to Eureka in 1885 to 1886 and secured a wealth of material. Photo courtesy of the Department of Entomology, National Museum of Natural History, Smithsonian Institution.

Although Colonel Casey made one of the greatest contributions to the taxonomy of the Coleoptera by any one man, much of his work has been soundly criticized. Many of his beetles have been determined by others to be synonymous with previously described species. Unlike his predecessors who used hand lenses to examine specimens, Casey used a binocular microscope with good light and was thus able to see details missed by others. He refused to consult other collections, preferring instead to examine specimens found in his own collection.

Casey frequently authored new species based on single specimens, distinguishing them from existing species by trivial details. For example, Casey described no less than 19 different species and several additional subspecies names for one of the largest beetles in California, the California Prionus *(Prionus californicus)*. He seldom took into account the variation among species populations. Many of his descriptions were of individual variants rather than valid species. Sadly, subsequent coleopterists have developed a general mistrust of Casey's publications.

Figure 6. Henry Clinton Fall (1862 to 1939) was California's first resident to make significant contributions to the study of the state's beetles. He assembled one of the finest private beetle collections in North America, representing 20,000 species. Photo from Linsley (1940), courtesy of the Pacific Coast Entomological Society.

Henry Clinton Fall (1862 to 1939)

Henry Clinton Fall (fig. 6) was California's first resident to make significant contributions to the study of the state's beetles, inspiring future generations of coleopterists, including E.C. Van Dyke and F. E. Blaisdell. Born in Farmington, New Hampshire, Fall received his bachelor's and honorary doctoral degrees in science from Dartmouth. Prompted by ill health, Fall moved to southern California in 1889 after a brief stint teaching high school mathematics in Chicago. He taught physical sciences at Pasadena High School, serving as the head of the science department there for nearly 25 years.

Inspired by a visit from G.H. Horn, Fall's first scientific article on beetles appeared in 1893. Subsequent papers included revisions of various beetle families and lists of species known from the California Channel Islands and southern California. Fall was one of the first researchers to work on insects, particularly Coleoptera, of the Channel Islands. After retiring from teaching in 1917, Fall moved to Tyngsboro, Massachusetts, where he continued to publish, curate, and identify specimens sent to him for determination. His 144th and last publication appeared in 1937.

At the time of his death, Fall had assembled what was up to that time the finest private beetle collection in North America. Reported to contain 200,000 specimens representing 20,000 species, his collection now resides appropriately next to those of LeConte and Horn in the Museum of Comparative Zoology at Harvard University.

Frank Ellsworth Blaisdell (1862 to 1947)

Born in Pittsfield, New Hampshire, Frank Ellsworth Blaisdell (fig. 7) moved with his family to San Francisco in 1870. The following year Blaisdell and his two younger brothers were stricken with scarlet fever, and only Frank survived. His poor health prompted his family to move to San Diego the same year. In 1875 the family moved to a ranch north of San Diego near Poway.

Largely self-taught and without the benefit of a modern high school or college education, Blaisdell enrolled in Cooper

Figure 7. Frank Ellsworth Blaisdell (1862 to 1947) grew up in San Diego County. After his retirement as a physician, Blaisdell worked on beetles at the California Academy of Sciences from 1924 to 1945, where his collection of some 200,000 beetles now resides. Photo from Van Dyke (1947), courtesy of the Pacific Coast Entomological Society.

Medical College, San Francisco, and received a doctor of medicine degree in 1889. He moved to San Diego to establish a medical practice but was unsuccessful. He returned to northern California in 1892, where he settled as a physician with his wife and son in the mining camp of Mokelumne Hill. In 1906 he began working at Cooper Medical College as a demonstrator and retired in 1927 as Professor of Surgery in charge of Surgical Pathology. After his retirement, Blaisdell worked on darkling (Tenebrionidae) and soft-winged flower (Melyridae) beetles at the California Academy of Sciences from 1924 to 1945, where his collection of some 200,000 beetles now resides.

Edwin Cooper Van Dyke (1869 to 1952)

Edwin Cooper Van Dyke (fig. 8) was California's first homegrown coleopterist. Born in Oakland, his family moved to Los Angeles in 1885, where his budding interest in botany and insects flourished. His first collecting trip was to Yosemite Valley in 1890, which he reached by pack train. Van Dyke earned his bachelor's degree in 1889 from the University of California, Berkeley, and his doctorate in medicine from Cooper Medical

Figure 8. Edwin Cooper Van Dyke (1869 to 1952) was an outstanding authority on the beetles of the Pacific States. He was also an expert on forest pests and the distribution of insects in North America. Photo from Essig (1953), courtesy of the Pacific Coast Entomological Society.

College in 1895. After serving as a physician in several hospitals in San Francisco and Baltimore, Van Dyke became the Physician in Charge at the Good Samaritan Mission in San Francisco from 1903 to 1912, a position he held while attending to his own medical practice. In 1913, Van Dyke returned to his first love, insects. He accepted the appointment of Assistant in Entomology at the University of California, Berkeley and became a full professor in 1927.

Van Dyke was an outstanding authority on the beetles of the Pacific States and made substantial contributions to the study of numerous families. He was also an expert on forest pests and the distribution of insects in North America. Van Dyke was an inspirational teacher whose door was always open to students. A tireless, enthusiastic, and conscientious collector, he was always mindful of preserving the habitat to ensure that collecting grounds remained fruitful for subsequent collectors. His collection of 200,000 beetles was presented to the California Academy of Sciences in 1924.

Hugh Bosdin Leech (1910 to 1990)

Born in Kamloops, British Columbia, Hugh Bosdin Leech's (fig. 9) early years were spent on the family farm in Salmon Arm, British Columbia. His interest in natural history, especially beetles, was apparent from an early age. He published his first paper in 1930, reporting a new host record for the longhorn beetle *(Phymatodes vulneratus)*. Leech earned his bachelor's degree from the University of British Columbia and master's degree from the University of California, Berkeley.

From 1930 to 1947 Leech worked in the Forest Entomology Laboratory in Vernon when not attending university. In 1947 he joined the staff of the California Academy of Sciences, where he became the Associate Curator of Coleoptera. Over the years he reorganized the beetle collection several times, processing tens of thousands of specimens each month from

Figure 9. Hugh Bosdin Leech (1910 to 1990) joined the staff of the California Academy of Sciences, where he became the Associate Curator of Coleoptera. He inspired many of today's coleopterists. Photo from Kavanaugh and Arnaud (1981), courtesy of the Pacific Coast Entomological Society.

collectors and shipping them on loan to researchers around the world. Leech reserved his days at the Academy for correspondence, curating, and visitors, conducting his research and writing in the evenings at home in "The Beetle Room." Leech was extremely influential and inspired many coleopterists living and working in California and elsewhere.

Earle Gorton Linsley (1910 to 2000)

Earle Gorton Linsley (fig. 10) was born in Oakland and developed his interest in insects as a child under the influence of E.C. Van Dyke and others. He attended the University of California at Berkeley, where he submitted his dissertation on the longhorn beetles of California. After graduation Linsley joined the Berkeley faculty as an instructor in entomology, became a full professor in 1953, and retired in 1973. He became one of the world's leading authorities on longhorns. Linsley's monumental work, *The Cerambycidae of North*

Figure 10. Earle Gorton Linsley (1910 to 2000) was one of the world's leading authorities on longhorn beetles (Cerambycidae). He was also an authority on rain beetles (Pleocomidae), as well as the biology, ecology, and taxonomy of native bees. Photo courtesy of John Chemsak.

America, completed in 1997 with J. A. Chemsak, stands as the only work to cover the entire American fauna. He was also an authority on the biology, ecology, and taxonomy of native bees. He published more than 400 books and articles. And, like several other California coleopterists, Linsley's first, and most enduring, entomological pursuit was rain beetles. He described 12 species or subspecies and established the taxonomic foundation for the group.

The Future of Beetle Studies in California

The history of beetle study in California is still being written. In spite of more than 200 years of exploration, many of California's regions remain poorly sampled or unexplored. As museum budgets continue to shrink, the burden of collecting, cataloging, and describing California's beetle fauna will depend on close cooperation between dedicated amateurs and skilled professional entomologists. There is still much to be done.

CHAPTER 2
FORM, DIVERSITY, AND CLASSIFICATION

For more than 250 million years the compact and armored bodies of beetles have evolved and adapted to meet the demands of an ever-changing environment. Through trial and error beetles have overcome numerous obstacles to ensure their reproductive success. By virtue of their sometimes striking bodily modifications and curious behaviors, beetles have developed the ability to invade all kinds of plant tissues, burrow through seemingly impenetrable soils, navigate fast-moving streams, or survive blistering deserts with aplomb.

Biting mouthparts, protective wing covers, and a developmental strategy that minimizes competition between adults and their offspring equip beetles to succeed in a dizzying array of terrestrial and freshwater habitats. And despite their bulkiness, beetles are quite capable, albeit clumsily at times, of flight. Described as "airborne trucks in low gear," most beetles manage to take to the air in search of food, mates, and new territories to exploit and conquer.

Armed with a basic understanding of the beetle body, we can begin to appreciate just how these amazing animals became one of nature's most successful groups of organisms. The modifications of their bodies not only provide the foundation for their classification, but also offer clues that reveal their evolutionary history and the reasons for their unparalleled success.

Dressed for Success: External Morphology

Unlike animals with backbones, beetles wear their skeletons on the outside of their body. This external covering, or *exoskeleton,* can be thought of as a highly modified cylinder that functions both as skeleton and skin. The exoskeleton not only protects the beetle, it also serves as the beetle's window to the world, providing a platform for important structures sensitive to touch and specific chemical compounds.

It's a Wrap: The Exoskeleton

A beetle's exoskeleton is composed of a polysaccharide called *chitin.* Chitin combines with the protein *sclerotin* to create a multilayered exoskeleton with the basic properties of plastic, tough yet light and flexible.

The exoskeleton is subdivided into small plates, or *sclerites.* The sclerites are fused together to form distinct body regions and their appendages. The sclerites are joined together by membranes of pure chitin or are separated by grooves called *sutures.* Examples of sclerites are those seen on the underside of the beetle belly, which looks like a series of plates. Beetle bodies are divided into three basic body regions: the head, thorax, and abdomen (fig. 11).

The division of the exoskeleton into body regions and sclerites allows for greater flexibility, much like the joints of a

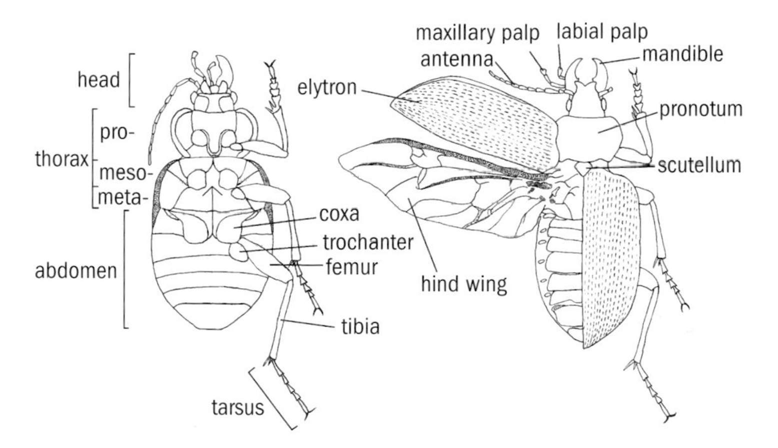

Figure 11. Ventral and dorsal view of the Common Black Calosoma (*Calosoma semilaeve,* Carabidae) illustrating the major body parts of an adult beetle.

knight's armor. But the exoskeleton does not just cover the outside of the beetle. It also penetrates the body, forming a complex system of internal supports that provide attachments for the powerful muscles that drive the mouthparts, legs, and wings.

Scattered about the outer layer of the exoskeleton, or *cuticle,* are bristly or flattened *setae.* Sometimes hairlike, the setae are sensory in nature, connecting the beetle to its environment by transmitting tactile and odor information directly to the nervous system. Strategically placed patches of dense setae also protect burrowing beetles from abrasion as they bore through wood or soil. When located between moving body parts, these tufts function as a dry lubricant or buffer to reduce wear and tear. A waterproof coat of wax further protects the cuticle. Secreted by glands embedded in the lower layers of the exoskeleton, the waxy layer helps beetles to retain body moisture.

The Desert Ironclad Beetle (*Asbolus verrucosus,* Tenebrionidae), has a tough exoskeleton that suggests the roughened exterior of the impenetrable metallic warships of the American Civil War. During periods of low humidity, its exoskeleton is coated with a waxy bloom that gives the beetle an almost ghostlike bluish hue. The wax filaments are secreted by glands whose spigots open at the tips of the knoblike bumps on the wing covers. The filaments spread over the upper surface of the body, creating a meshwork that functions as a deep-sea-diving suit in reverse, sealing off the beetle from its dry surroundings to keep water inside the body.

The thickness and durability of the beetle exoskeleton also affords them some protection from predators and other potentially life-threatening predicaments. For example, adult ant-loving scarab beetles (*Cremastocheilus* spp., Scarabaeidae) are predators that feed on ant larvae. Ant colonies can be particularly hostile environments for uninvited guests, especially those dining on their brood. Ant-loving scarabs are cleverly adapted for dealing with their pugnacious hosts. Their heavily armored exoskeleton is fitted with special plates and grooves, allowing the beetle to safely tuck away its sensitive antennae and delicate mouthparts from the snapping jaws of the ants. The surface of the beetle's exoskeleton resembles the nooks and crannies of an English muffin. Like a sponge, the scarab's exoskeleton quickly absorbs the characteristic odor of the ant nest where it lives only temporarily. As soon as the ant-loving scarab smells like the rest of the nest, it is accepted as part of the family and can hunt for larvae in the nest with impunity. The ruse is further backed up by special glands on the beetle's thorax that produce appeasement substances that have a calming effect on belligerent ants.

The Path of Least Resistance: Reducing Drag

Flattened from top to bottom and with smooth, uninterrupted outlines all around, the bodies of whirligigs (Gyrinidae), predaceous diving beetles (Dytiscidae), and water scavengers (Hydrophilidae) are wonderfully *hydrodynamic.* The streamlined bodies of these aquatic beetles allow them to glide through the water with ease.

Adult wood-boring beetles are also sleek, effectively reducing the amount of energy they must spend chewing through wood. The projectile-like bodies of metallic wood-boring beetles (Buprestidae) and the stout cylindrical body plan of some longhorn beetles (Cerambycidae) require an excavation relatively small in diameter as they chew their way through solid wood.

The bodies of some predatory beetles are extremely flattened, an adaptation allowing them to fit into tight spaces. The low profile of ground beetles (Carabidae) enables them to hunt for food among the narrow tangles of vegetation. The fire engine red Flat Bark Beetle (*Cucujus clavipes puniceus,*

Plate 2. Adult and larval Flat Bark Beetles (*Cucujus clavipes puniceus*, Cucujidae) (10 to 17 mm), prey on bark and other wood-boring beetles. Both are capable of surviving extended freezing temperatures.

Cucujidae) (pl. 2), and its larvae are so flat that they, in pursuit of insect prey, can fit neatly in the narrowest of spaces between the bark and wood of dead trees.

The Head

The beetle head is usually the first part of the body to come into contact with the environment and is therefore packed with important structures. Capsulelike in shape, the head is sturdily reinforced internally and attached to the body by a flexible, membranous neck. Sometimes the neck is quite visible, as in the blister beetles (Meloidae), whose antlike heads are completely exposed. In most beetles the neck is hidden and the head is partially nested inside the body. In fireflies and glowworms (Lampyridae), the head is partially or wholly concealed from above by part of the thorax.

The Eyes

The round and occasionally bulging compound eyes are usually broadly separated from each other. They are composed of dozens or hundreds of individual facets or lenses, each composed of colorless cuticle. In addition to visible light, beetles can perceive ultraviolet and even infrared radiation, but just how they view the world is still open for debate because the resultant image is dependent upon how their brain processes it. Most beetles probably view the world as a single image made up of many pieces, as opposed to the mosaic of separate images popularized by children's toys.

Awash in light, the lenses of day-active beetles are relatively small and flat. Nocturnal species possess more convex lenses that gather as much available light as possible. Small compound eyes with fewer lenses are common in flightless or nocturnal species. Beetles living in caves or that dwell deep in leaf litter live in total darkness and may lack eyes altogether.

The compound eyes of beetles may be partially or completely divided by a ridge of cuticle called the *canthus*. The eyes of whirligigs are completely divided by the canthus. The exposed upper half of the eyes possess lenses specifically adapted for gathering images in the air, whereas the submerged lower half is adapted for collecting images in a thick and watery medium.

Horned Heads

Beetle horns have evolved in several thousand species of beetles, appearing as long spikes, scooped blades, curved spines, blunt knobs, or tiny bumps, especially in males. In California, the Rugose Stag Beetle (*Sinodendron rugosum,* Lucanidae) (pl. 3), has a lone, short spike, whereas that of the earth-boring scarab (*Bolboceras obesus,* Geotrupidae) is long, slender, and curved backward. Another earth-boring scarab *(Ceratophyus gopherinus)* has a single pair of short,

Plate 3. The male Rugose Stag Beetle (*Sinodendron rugosum,* Lucanidae) (11 to 18 mm) has a lone, short spike on its head. The female, shown on the left, does not have a spike. It lives in isolated wet canyons of the Transverse and Peninsular Ranges in southern California and throughout the wooded areas of northern and central California northward to British Columbia, where it is sometimes common beneath the bark of wet, rotten logs of hardwoods.

opposing horns, one mounted on the head and one on the pronotum.

Despite the diversity of horn shapes, sizes, and the beetles that use them, most horns seem to be used for the same basic purpose: to battle with other males for resources that attract females. For example, the male of the Sleeper's Elephant Beetle (*Megasoma sleeperi,* Scarabaeidae) bears a single forked horn on its head. Male elephant beetles probably use their horns to defend sapping spots on palo verde trees *(Cercidium floridum)* from other males as they wait for the arrival of hungry and sexually receptive females.

Bigger Is Not Necessarily Better

The variable development of horn size in male beetles is of particular interest to scientists who study mate selection. What are the factors that affect horn size in the first place? Environmental factors, much more than heredity, play an important role in the development of beetle horns in the larval stage. Proper nutrition is critical. In the dung beetle (*Onthophagus taurus,* Scarabaeidae), once the larvae crosses a certain size threshold the presence of key developmental hormones trigger the growth of big horns. Smaller male larvae that fail to reach this critical threshold fail to develop horns, even in the presence of the developmental hormones. The study of this phenomenon where the growth rates of some body parts, like horns or mouthparts, are accelerated in relation to overall size is called *allometry.*

But is there a distinct advantage to horn size? Do well-endowed males always get the female, at the expense of their lesser-equipped counterparts? *Onthophagus taurus* has two distinct sets of males based on size and horn development. Both classes of beetles have evolved successful, yet different, strategies for mating.

Females burrow beneath dung to feed themselves and stockpile food for their young. Males must gain access to the females via the burrow. Well-endowed males guard the entrances to the female's burrow, using their horns to drive off rival suitors. In this world, size matters because the larger, more powerful males nearly always overcome smaller rivals, with or without horns. Rather than risking certain failure by approaching the more powerful horned male head-on, hornless males instead sneak into the female's burrow, slipping past the male by digging a side tunnel that crosses the heavily defended burrow of the female.

Oral Fixation: The Mouth

Beetles possess chewing mouthparts variously modified to cut flesh (e.g., ground beetles and tiger beetles [Carabidae]), grind leaves (e.g., leaf beetles [Chrysomelidae]), or strain fluids (dung beetles [Scarabaeidae]). Conspicuously large or otherwise, the structure of the mouthparts of all beetles is founded on the same basic plan: an upper lip, or *labrum;* two pairs of chewing appendages, the *mandibles* and *maxillae;* and a lower lip, or *labium.*

Beetle mandibles often serve as multipurpose tools. The Pine Sawyer (*Ergates spiculatus spiculatus,* Cerambycidae) (pl. 4) wields its mandibles as implements for burrowing through wood and as defensive weapons. Male tiger beetles of the genus *Cicindela* use their mandibles to tear apart insect prey and as sexual restraints during copulation.

The maxilla and labium usually possess delicate, flexible structures, or *palps,* that function like fingers to manipulate food and shove it into the mouth. The maxillary palps of water scavengers are quite conspicuous and are nearly as long as the antennae.

Protecting the mouthparts from above is a broad plate of cuticle formed by the leading edge of the head known as the

Plate 4. The Pine Sawyer (*Ergates spiculatus spiculatus,* Cerambycidae) (40 to 65 mm), the largest beetle in California, is armed with powerful jaws for gnawing through wood.

clypeus. Below the head and behind the mouthparts are two sclerites known as the *mentum* and the *gula.*

The mouthparts of some beetles are *prognathous,* directed forward and parallel to the long axis of the body. Prognathous mouthparts are typical of predators such as ground beetles and whirligigs, as well as some wood-boring beetles (e.g., some prionine Cerambycidae). *Hypognathous* mouthparts are directed downward and are typical of most plant feeders, including chafers (Scarabaeidae), many longhorn beetles (Cerambycidae), leaf beetles, and weevils (Curculionidae). The hypognathous mouthparts of net-winged beetles (Lycidae) and many weevils are drawn out into a beak.

Smelly Feely: The Antennae

Lacking a nose, beetles must rely on their antennae for their sense of smell. Each *antenna* is mounted on the head just between the base of each mandible and compound eye. Equipped with an array of incredibly sophisticated receptors, the antennae are used to detect food, locate mates and egg-laying sites, and assess environmental conditions. Ground beetles and rove beetles (Staphylinidae) possess specialized comblike structures on their legs and feet used to keep their antennae scrupulously clean. Although primarily organs of smell, the antennae are also quite sensitive to touch.

The antennae have other practical purposes. Rising to the surface of the water head first to replenish their air supply, water scavenger beetles break through the surface film with the aid of their antennae. Some male blister beetles use the antennae of their female partners as handles during copulation, grasping them with their mandibles.

Each beetle antenna consists of three basic parts: *scape, pedicel,* and *flagellum* (fig. 12). In whirligigs the scape and pedicel contain a highly developed group of sensory cells called *Johnston's organ.* Johnston's organ is believed to func-

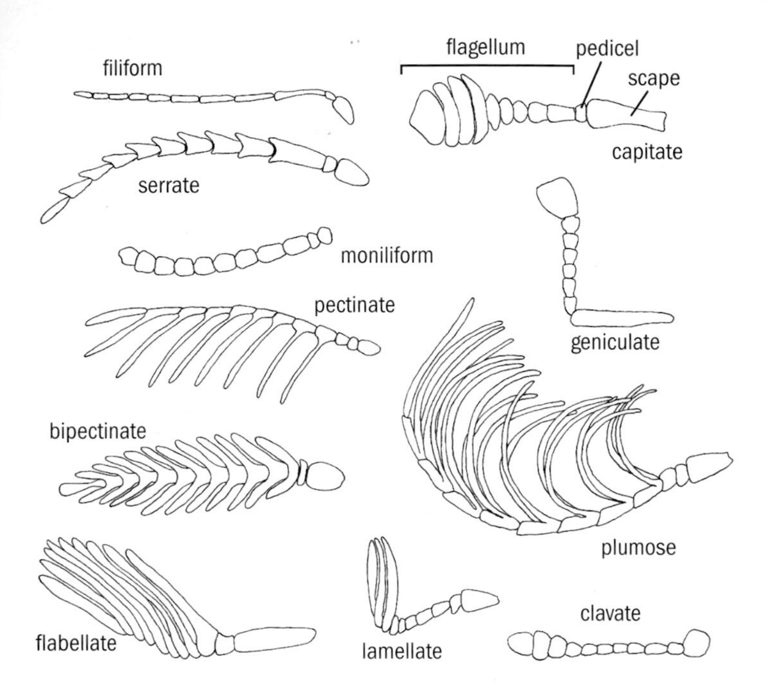

Figure 12. Examples of types of antennae possessed by California beetles: Filiform, or threadlike (Oregon Tiger Beetle [*Cicindela oregona*], Carabidae); capitate, or clubbed (Black Burying Beetle [*Nicrophorus nigrita*], Silphidae); serrate, or saw-toothed (California Prionus [*Prionus californicus*], Cerambycidae); moniliform, or beadlike (*Nyctoporis carinata,* Tenebrionidae); geniculate, or elbowed (Yucca Weevil [*Scyphophorus yuccae*], Curculionidae); pectinate (*Euthysanius lautus,* Elateridae); bipectinate (*Pleotomus nigripennis,* Lampyridae); plumose, or feathery (California Phengodid [*Zarhipis integripennis*], Phengodidae); flabellate (*Sandalus* sp., Rhipiceridae); lamellate (Dusty June Beetle [*Parathyce palpalis*], Scarabaeidae); clavate (*Ostoma pippingskoeldi,* Trogossitidae).

tion as part of an echolocation system used to detect vibrations on the water surface. Surface vibrations emanate from the movement of the whirligig's predators and prey, as well as from other whirligigs.

Although the antennae appear to be segmented, they lack internal musculature. Without muscles, the segments are more correctly called *antennameres.* The usual number of antennameres for beetles is 11, but 10 or fewer antennameres are common.

The antennae exhibit an amazing array of modifications that are used to classify beetle families (fig. 12). They are generally shorter than the body and somewhat similar in both sexes; however, the antennae of some groups differ considerably between males and females. For example, the male Oregon Fir Sawyer (*Monochamus scutellatus oregonensis,* Cerambycidae) has long, threadlike antennae that are up to three times the length of the body, whereas the antennae of the female are only slightly longer than the body. Ornate modifications are especially evident in species that track airborne chemical attractants called *pheromones* that advertise the sexual readiness of otherwise secretive females. Males of some beetles (e.g., rain beetles [Pleocomidae], dung beetles, phengodid beetles [Phengodidae], longhorn beetles) often have elaborate antennal structures resembling fans or feathers to increase the surface area available for sensory pits. These pits are capable of detecting even the smallest amounts of these stimulating chemical messengers.

The Powerhouse of the Body: The Thorax

The thorax encloses strong muscles that drive the legs and wings. It is divided into three ringlike segments: the *prothorax* nearest the head, the *mesothorax,* and the *metathorax* (fig. 11). A pair of legs is attached to each segment. Only the mesothorax and metathorax possess wings.

The prothorax is the conspicuous midsection of the beetle body. The remaining two thoracic segments are firmly joined together and are concealed beneath the wing covers. The lateral or side margins of the prothorax are variable. In aquatic families the prothoracic margins are sharply keeled to help streamline and stabilize their body in the water. Slowly rambling along the ground or climbing on their food plants, blister beetles have evolved prothoracic margins that are more or less rounded from top to bottom.

The top sclerite of the prothorax is called the *pronotum* and is variable in shape and texture. In the antlike flower beetles (Anthicidae) the pronotum is hornlike, extending forward over the head. The pronota of some earth-boring beetles and twig borers (Bostrichidae) are scooped out like a bulldozer and are used to clear their excavations of soil and sawdust. The sculpturing of the pronotum—such as the size and placement of horns, pits, and bumps—is often useful in the identification of beetle species.

One of the best-known modifications of the beetle thorax is that of the click beetles (Elateridae). Lying on their backs, click beetles right themselves by flipping into the air with an audible click. They accomplish this feat by arching their bodies backward so only the prothorax and wing covers are in contact with the ground. By quickly contracting their muscles, a stout spine on the prothorax catches on the edge of a groove located on the mesothorax. As the muscle tension builds, the spine suddenly snaps into the cavity. The abrupt release of energy jerks the beetle into the air and, with luck, it lands in an upright position, out of harm's way, where it can then scurry to safety.

Legs

Beetle legs are subdivided into six parts (fig. 11). The *trochantin* is partially or completely hidden by the *coxa* when the leg is in normal position. The coxae firmly anchor the legs into the coxal cavities of the thorax yet allow for the horizontal to-and-fro movement of the legs. Usually small, the coxae of ground beetles and their relatives extend backward across part of their abdomen, an important diagnostic feature. In the crawling water beetles (Haliplidae) the coxae form broad plates that conceal nearly the entire abdomen. The next three segments are the *femur* (thigh), *tibia* (shin), and *tarsus* (foot).

Each tarsus (plural *tarsi*) consists of four or five segments, sometimes fewer. Like the antennae, the tarsi lack any sort of musculature, and their segments are technically called *tarsomeres.* The number of segments on each foot is an important clue as to the identity of beetles. The tarsal formula used in descriptions of beetles, such as 5-5-5, 5-5-4, or 4-4-4, indicates the number of tarsomeres on the front, middle, and hind legs, respectively. Some segments are difficult to see without careful examination under high magnification. The tarsus usually bears a pair of claws attached to the last segment. Claws lacking notches, teeth, or saw-toothed edges are referred to as "simple."

The front tarsi of some male predaceous diving beetles (e.g., *Cybister* and *Dytiscus*) are highly modified. They function like suction cups, enabling the male to grasp the smooth and slippery wing covers of the female while mating. The feet of some longhorns and leaf beetles are equipped with broad, sticky pads made up of tightly packed setae, enabling them to walk on smooth vertical surfaces or cling to slippery or uncooperative mates.

The legs of beetles are variously adapted for living on land or in the water. The long, slender legs of ground beetles and tiger beetles are modified for running after prey, whereas the thick, rakelike legs of scarabs are used for digging. The last two pairs of legs of whirligigs are short and paddlelike, whereas those of predaceous diving beetles and water scavengers are long and fringed with setae. These oarlike legs propel them through the water with amazing speed and dexterity.

Wings

The most prominent and characteristic feature of nearly all beetles is their thick and leathery wing covers, known collectively as *elytra* (fig. 11). Separately each wing cover is called an *elytron.* The elytra meet along the *elytral suture,* the distinctive line that runs down the middle of their back. At the base of the elytra, located on the suture, is a small, round or triangular sclerite that may or may not be visible called the *scutellum.*

The elytra cover most or all of the abdominal segments but are typically short in the rove beetles, sap beetles (Nitidulidae), histerids (Histeridae), soft-winged flower beetles (Melyridae), phengodids, and ripiphorids (Ripiphoridae). Some longhorn beetles, blister beetles, and click beetles also have short elytra. Adult females of both phengodids and glowworms lack wings altogether and resemble larvae.

The elytra help stabilize beetles in flight. In most species the elytra are lifted and separated when airborne, but in some fast flyers, such as the Green Fruit Beetle (*Cotinis mutabilis,* Scarabaeidae) and metallic wood-boring beetles (e.g., *Acmaeodera,* Buprestidae) the elytra are partially or totally fused along the elytral suture. At lift off, the fused elytra lift slightly as the flight wings unfurl from the sides.

The elytra also protect the delicate membranous flight wings and internal organs. Elytra effectively conserve body fluids and shield beetles from extreme temperatures. The *subelytral space,* located between the elytra and abdomen, insulates desert species, such as the wingless Inflated Blister Beetle (*Cysteodemus armatus,* Meloidae) (pl. 5), from high temperatures and minimizes water loss through respiration. In desert darkling beetles (Tenebrionidae) the subelytral cavity functions as a convective-cooling system, drawing heat away from the body. It also acts as a heat-buffering system by interjecting an insulating layer of air between the beetle's

Plate 5. The subelytral space of the wingless Inflated Blister Beetle (*Cysteodemus armatus,* Meloidae) (7 to 18 mm) protects it from high temperatures and minimizes water loss through respiration. It is found in the Mojave and Colorado Deserts in spring, actively crawling on flowers or the ground during the warmer parts of the day.

body and the hot desert sun. This same cavity also serves as a storage space for air bubbles, enabling aquatic beetles to carry their air supply with them so they can remain underwater for extended periods. The effectiveness of this cavity in desert beetles is sometimes enhanced by the fusion of the elytral suture.

In the Western Pine Beetle (*Dendroctonus brevicomis,* Curculionidae), the elytra of the males are adapted for *stridulation,* or sound production. Ridges along the elytral suture, the line separating the elytra, are rubbed together to produce clearly audible sounds that alert the female to the male's presence as he enters the gallery in the bark.

A framework of veins supports the membranous hind wings. Some of these veins are hinged, so the wings can be carefully folded and tucked underneath the elytra. The flight wings of a few desert blister beetles, many darkling beetles, and some weevils are reduced in size or absent altogether. In rain beetles and some scarabs the males are fully winged, whereas the females lack functional hind wings.

Gasteronomic Review: The Abdomen

Each abdominal segment is a ring made up of four sclerites: a dorsal *tergum* (plural *terga*), the ventral *sternum* (plural *sterna*), and two lateral *pleura* (singular *pleuron*) (fig. 11). In beetles whose abdomens are completely covered by the elytra, the terga are thin and flexible. In rove beetles, with their short elytra and exposed abdomens, the terga are thick and rigid. Adults usually have nine abdominal segments, but only the first five or six segments are usually visible. Housed within the abdomen is the bulk of the circulatory, respiratory, digestive, excretory, and reproductive systems.

The abdomen also performs other important tasks. The flattened and narrow whirligig abdomen flexes sharply downward and side to side and is used as a rudder. This gives whirligig beetles the ability to gyrate wildly on the surface of ponds, lakes, and slow-moving streams. Upon landing, rove beetles use the tips of their abdomens to carefully tuck their long flight wings beneath their short elytra.

The last three or four abdominal segments are generally not visible and are variously modified for reproductive activities: egg laying in females and copulation in males. The female egg-laying tube, or *ovipositor,* is kept internally. Long ovipositors are characteristic of beetles that deposit their eggs deep in sand or plant tissues. Short and stout ovipositors are used to deposit eggs directly on the surface of various substrates. The copulatory organs of males, often adorned with various hooks and spines, are used to grasp and hold the female internally while mating. These elaborate reproductive organs are often of considerable value when identifying species of beetles.

It's What's Inside That Counts: Internal Anatomy

Beetles must do the same things that all other animals do: sense their environment, digest food and excrete waste, circulate blood, take in oxygen and expel carbon dioxide, and reproduce. Despite their small size they have evolved complex internal organ systems that allow them to function effectively in the environment.

A Gaggle of Ganglia: The Nervous System

The nervous system consists of a pair of ventral nerve cords and nerve bundles *(ganglia)* that conduct nerve impulses to all parts of the beetle's body. Within the head lies a three-part brain. The *protocerebrum* processes images coming in through the optic nerve, whereas the *deuterocerebrum* controls and interprets information from the antennae. The paired *tritocerebral lobes* control the labrum and the foregut. The *subesophageal ganglion* controls the remaining mouthparts, whereas the *thoracic ganglia* process nerve impulses that activate the legs and wings. The separate, mobile nature of the prothorax in beetles is mirrored internally by a large and distinct *prothoracic ganglion.* The remaining two thoracic ganglia may be more or less fused and are followed by a series of abdominal ganglia.

Gutsy Moves: The Digestive System

Solid and liquid foods processed by the mouthparts are passed into the digestive system, a conveyor belt that moves materials from the mouth to the anus. Along the way water is extracted and nutrients are absorbed. Solid materials are broken down by a battery of grinding teeth in the *proventriculus,* located at the rear of the *foregut.* Digestive enzymes secreted by the salivary glands work in the *midgut* to reduce the food into its most basic components before it can be absorbed as nutrients. The *hindgut* is the primary site of water absorption. Here the waste is converted into dry, often distinctive, fecal pellets known in polite company as *frass.* Frass is eliminated from the body through the *anus.*

But waste not, want not, as some beetles recycle their own excrement. The larvae of some tortoise beetles (Chrysomelidae) camouflage themselves in a protective layer of their own frass in an attempt to look unappetizing. Grubs of the Green Fig Beetle *(Cotinis mutabilis)* construct pupal chambers using their own frass and saliva like bricks and mortar. The hardened walls of the completed egg-shaped chamber prevent the pupa from becoming dehydrated and deter predators.

In addition to the elimination of solid waste generated by the digestive tract, beetles remove waste from their blood via the *Malpighian tubules,* named after the Italian scientist Marcello Malpighi, who first described these organs in silkworms in 1669. The Malpighian tubules extract uric acid and salts, dumping them into the hindgut, where any remaining water and some salts are absorbed, leaving uric acid crystals to be discharged with the frass.

You Gotta Have Hearts: The Circulatory System

Unlike the closed system of vertebrates where the blood remains inside arteries and veins, beetles possess an *open circulatory system.* The primary components of this system include a *dorsal aorta* fitted with a series of pumps, or *hearts,* found in the abdomen. Located between the hearts are the *ostia,* openings that allow the flow of blood from the body cavity back into the aorta. The dorsal aorta extends from the abdomen to the head. The pumping action of the muscles surrounding the hearts forces blood into the head, where it spills out into various body cavities, bathing the tissues in nutrient-rich blood. It also contains hormones that control the beetle's life cycle. Eventually the blood travels back into the abdomen, where it enters the aorta through the ostia and is again pumped forward to the head by the hearts.

Never Breathless: The Respiratory System

The basic respiratory system of beetles enables them to survive in both terrestrial and freshwater habitats. The system consists of a network of *tracheae,* or breathing tubes, and *air sacs.* The tracheae are supported externally by a series of coils, or *taenidia,* much like the ribs on vacuum cleaner hoses. The tracheae branch into progressively smaller tubes until they come into direct contact with the cells of the muscles and internal organs.

The air sacs are vital to the beetle's good health and are formed at the blind ends of the tracheae or by expansions along the main tracheal trunks. They keep the body light, help to dissipate body heat, maintain blood pressure, and improve tracheal ventilation by supplementing the opening and closing of the valves, or *spiracles.* Some beetles can partially close their spiracles as a water conservation measure, reducing the amount of body moisture lost through respiration.

Oxygen enters the tracheal system through spiracles, located on the sides of the thorax and abdomen. Carbon dioxide is expelled from the body through the same system. Without lungs, these respiratory gases move passively throughout the body by diffusion. Through the process of diffusion, high concentrations of oxygen and carbon dioxide invade regions with little or none of these respiratory gases. The dependence of beetles and other insects on this passive transport system is one reason why these animals remain small in size, because diffusion is most effective only across short distances.

Scuba Gear: Respiration in Water

Aquatic beetles must regularly bring fresh supplies of air into contact with their spiracles. Water scavengers do this by breaking through the surface tension headfirst with their antennae to draw a layer of air over the lower surface of their abdomen. Predaceous diving beetles trap air beneath their elytra by breaching the surface with the tip of their abdomen. In both cases the trapped bubble of air eventually functions as a *physical gill.* As the beetle consumes the oxygen of the bubble, the partial pressure of oxygen within the bubble falls below that of the dissolved oxygen in the surrounding water. Dissolved oxygen from the water diffuses into the bubble, but not quickly enough to prevent its depletion. As the oxygen supply is exhausted, the nonrespiratory gas nitrogen begins to diffuse out of the bubble. The loss of nitrogen reduces the size of the bubble and its ability to obtain dissolved oxygen from the water. When the bubble becomes too small, the beetle must return to the surface for more air. The longevity of the bubble is determined by the rate of consumption by the beetle, the bubble's surface area, and water temperature. The physical gill may also help control the beetle's buoyancy in water, much like the swim bladder in fish.

A modification of this system is called *plastron breathing.* The plastron is a film of air held in place by a thick layer of water-repellent setae. The film can partly or wholly envelop the beetle's velvety body underwater. Plastron breathing is not very effective for very active beetles and is used largely by sedentary grazers, such as the long-toed water beetles (Dryopidae) and riffle beetles (Elmidae).

Everything You Wanted to Know: Reproduction

The male reproductive system consists of a pair of *testes* whose ducts, the *vasa differentia,* join to form the ejaculatory duct. The female reproductive organs roughly mirror those of the male and consist of a pair of *ovaries,* each with one or more *ovarioles.* In dung beetles such as *Canthon* and *Onthophagus,* only the left ovary is present. Short-lived beetles generally tend to have more ovarioles, increasing the number of eggs available at any given time to maximize their reproductive potential. Long-lived beetles do not need to produce all their eggs at once and generally have fewer ovarioles.

Each ovary is equipped with an *oviduct* that leads into a common *median oviduct.* Opening into the median oviduct is the *spermatheca,* a blind sac where live sperm are stored for up to several months. The eggs are fertilized as they pass through the median oviduct and then laid. The haphazard or systematic placement of the eggs is called *oviposition.* The median oviduct or other associated glands secrete a sticky coating over the eggs, causing them to adhere to one another or to attract a protective coating of sand or debris. For example, the Cactus Longhorn Beetle (*Moneilema semipunctata,* Cerambycidae) lays its eggs on the ground near the base of a cactus. Covered with sticky glue, the eggs soon become coated with a protective layer of sand. Female Giant Water Scavenger Beetles (*Hydrophilus triangularis,* Hydrophilidae) set their eggs adrift in a silken cocoon that functions as both floatation device and sail.

Youthful Appearances: Larval Morphology

As is typical with all insects that undergo complete metamorphosis, the beetle larva bears absolutely no resemblance to the adult. The larval head sometimes has a Y-shaped groove, or *ecdysial suture,* a line of predetermined weakness in the exoskeleton. The ecdysial suture functions as part of a hatch through which the larva can escape its old larval exoskeleton during the molting process. The larvae lack compound eyes, but most possess from one to six simple eyes, or *ocelli,* on each

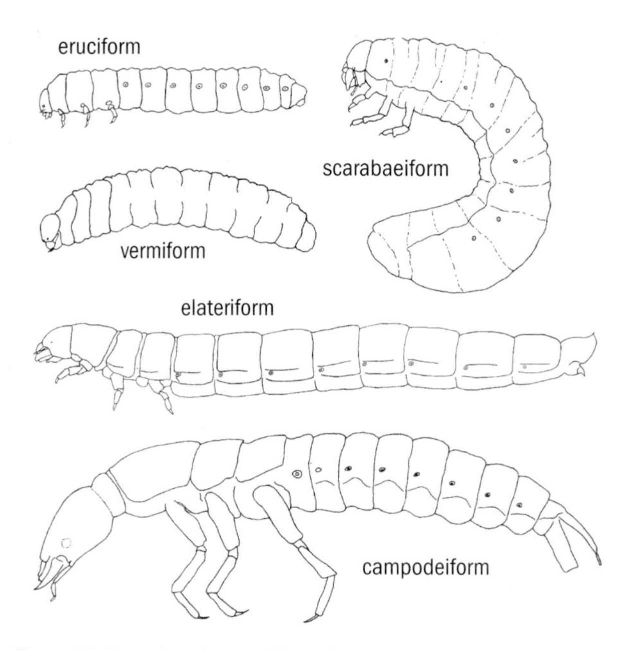

Figure 13. Examples of types of larval body forms: Eruciform or caterpillar-like (Chrysomelidae); Vermiform or maggot-like (Curculionidae); Scarabaeiform or C-shaped (Scarabaeidae); Elateriform (Tenebrionidae); Campodeiform (Staphylinidae).

side of the head; the larvae of a few species lack eyes altogether. As in the adults, the mouthparts of larvae crush, grind, or tear foodstuffs. The antennae are not nearly as variable as in adult beetles and are usually of only two to four segments. The larvae of the Giant Water Scavenger possess sharp, pointed antennae that actually work with the mandibles to puncture the exoskeleton of insect prey. Most predatory larvae are fluid feeders that pierce and drain their victims.

The three thoracic segments are very similar to one another, but the one nearest the head may have a thickened plate on its back. The legs, when present, have six or fewer segments. The abdomen is generally nine or 10 segmented and is

Plate 6. The slow caterpillar-like grubs of ladybugs (*Hippodamia,* Coccinellidae) typify eruciform larvae. They possess well-developed heads, legs, and fleshy abdominal protuberances. Photo by C.L. Hogue.

usually soft and pliable, a feature no doubt useful for rapidly growing animals. The abdomen lacks legs, but some segments may be variously endowed with fleshy protuberances resembling warts. These structures help the larva to gain purchase as it moves about. The last abdominal segment may possess a pair of fixed or segmented projections called *urogomphi.*

Although the larvae of beetles vary considerably in form, at least five basic body types have been recognized (fig. 13). The slow caterpillar-like grubs of ladybugs (Coccinellidae) (pl. 6) and some leaf beetles typify *eruciform* larvae. They possess well-developed heads, legs, and fleshy abdominal protuberances. The *scarabaeiform* larvae of scarab and stag beetles (Lucanidae) (pl. 7) are sluggish, C-shaped grubs with distinct heads and well-developed legs. These grubs are perfectly suited for a subterranean life, buried out of sight in soil, decomposing plant material, or soft, punky wood. The highly ac-

Plate 7. Scarabaeiform larvae of the Rugose Stag Beetle *(Sinodendron rugosum)* are sluggish, C-shaped grubs with distinct heads and well-developed legs and are perfectly suited for life buried out of sight in soft, punky wood.

tive and agile larvae of predatory and parasitic beetles, such as ground beetles and the *triungulins* of blister beetles, are referred to as *campodeiform*. Their flattened, elongate bodies are equipped with well-developed legs and antennae. *Elateriform* larvae have long, straight, cylindrical bodies with tough exoskeletons and short legs. Both wireworms (Elateridae) and the Mealworm *(Tenebrio molitor)* are typical examples of this larval body form. As their name suggests, the *vermiform* larvae of some weevils are maggotlike, with thick and legless bodies.

Transitional Packaging: Pupal Morphology

During the pupal stage, the first physical details of the adult beetles are revealed. Unlike the pupae of butterflies and

Plate 8. The legs, wings, mandibles, and antennae are clearly visible and are loosely attached to the body of this scarab chafer pupa, the Dusty June Beetle *(Parathyce palpalis)*. Photo by C. L. Hogue.

moths where the wings and legs are firmly attached to the body, these appendages on the beetle pupa usually stick out. This type of pupa, where the appendages are clearly visible and are not tightly attached to the body, is known as *exarate* (pl. 8). The pupal appendages of ladybugs, leaf beetles, and some rove beetles, however, are tightly pressed against the body throughout their entire length.

The pupal abdomen may possess functional muscles, allowing for some movement. *Gin-traps,* present in some species, consist of specialized teeth located on opposing surfaces of the abdominal segments. Gin-traps can quickly snap shut and are thought to discourage attacks by small predators and parasitic mites.

The Age of Beetles

For almost two-and-a-half centuries, entomologists have labored to describe approximately 350,000 kinds of beetles in the scientific literature, more than for any other group of organisms. That's an average of slightly more than four species a day since 1758, the year we began to systematically keep track of such things. If one example of every plant and animal were placed in a row, every fifth species would be a beetle and every tenth species would be a weevil. With perhaps 8,000 or more species living in the state, beetles represent the largest and most diverse group of any organisms known in California. But how did they achieve such phenomenal success? Two factors have contributed to the overall triumph of beetles.

Nature's Most Successful Design

First, beetles are generally small and compact, permitting them to burrow, hide, search for food, find mates, and lay their eggs in a staggering array of environments, from the steadily shifting sands of coastal beaches and deserts, to the rocky barrens high in the mountains. Their tough elytra protect them from abrasion as they bore through wood or burrow into soil and decomposing plants. The subelytral cavity of beetles predisposes them to exploit both desert and aquatic habitats with equal success. And, though not particularly fast or agile fliers, by taking to the air, beetles manage to avoid predators, locate mates, find food, and colonize new habitats.

Dances with Plants

Secondly, an evolutionary tango between beetles and plants has contributed significantly to the diversification of beetles. Based on fossil evidence, the earliest-known beetles appeared more than 230 million years ago and are believed to have consumed dead organic material and fungi. Additional evidence indicates that plant feeding among beetles arose some 50 million years later, about the time of the appearance of cycads, ginkgoes, pines, and other conifers.

As the dinosaurs began to vanish late in the Cretaceous period, about 65 million years ago, the first flowering plants, or *angiosperms*, appeared. Beetles were primed and ready to exploit these new and flowery plants. Already adept at using the vegetative structures of nonflowering plants, the shift from gymnosperms to angiosperms was relatively simple. Roots, stems, leaves, flowers, and fruits not only provided new sources of food for both adults and larvae, they also afforded novel sites for shelter and oviposition. The increased ecological opportunities created by the appearance of angiosperms offered beetles what was then a wealth of untapped resources ripe for the taking.

Angiosperms, however, are hardly passive partners in this coevolutionary relationship. Many plants responded to the plant-feeding activities of beetles and other animals by evolving prickly spines, bristly hairs, and sticky or toxic sap. Not so easily deterred, beetles have evolved strategies to circumvent or minimize these plant defenses. This ancient dance, conducted over millions of years, has put both angiosperms and beetles on an evolutionary fast track, resulting in the unprecedented diversity of both groups.

The Name Game

Carolus Linnaeus, a Swedish physician born early in the eighteenth century, fully understood the importance of carefully naming living things. He took on a job of biblical proportion by setting out to systematically describe all of the world's organisms. In the process of describing nearly 15,000 species of plants and animals, including about 2,000 insects, Linnaeus established the system of biological classification still in use today. He was also the first person to compare the anatomical features of plants and animals, a method considered at the time to be a major breakthrough.

Linnaeus also streamlined the actual naming of organisms. Previously, the names of plants and animals consisted of cumbersome strings of Greek and Latin descriptions. Linneaus continued to use descriptive Greek and Latin terms, but limited the number of words used to name each organism to just two: the *genus* (plural *genera*) and *species* (singular and plural). The genus and species combine to form the scientific name of an organism. Scientific names of the genus and species are always written in italics or are underlined.

The tenth edition of Linnaeus's *Systema Naturae* in 1758 marks the beginning of the Linnaean system of binomial nomenclature for zoology. The use of this universally recognized system greatly facilitates the storage and retrieval of biological information. Once properly identified, the scientific name of a beetle, or any organism, effectively unlocks the door to all the information available on that species.

The Species Concept

Our scientific efforts at classification focus on the *species*. A beetle species is generally considered to be a group of interbreeding individuals capable of producing successive generations of reproductively viable offspring. Each beetle species has a unique evolutionary history and inhabits a special biological niche within a particular geographical range. *Subspecies* are isolated populations of beetles capable of interbreeding with other isolated populations of the same species.

Name That Beetle

The scientific community only recognizes beetle species that have been formally described in a scientific publication. Beetles are assigned scientific names, usually of Latin or Greek derivation. The International Code of Zoological Nomenclature establishes procedures for affixing scientific names to all animals.

Unlike common names, which can vary considerably among regions, cultures, and languages, scientific names are universally recognized. Papers published around the world in different languages all use the same Latinized scientific names for the same species. Efforts to standardize the common names of beetles are usually applied only to economically important species.

In catalogs and other scientific documents, a name and year often follow the scientific name. Consider, for example, *Coenonycha dimorpha* Evans, 1986. Evans was the first person to describe this species of scarab beetle and formally published it in 1986. The year establishes the priority of the name *Coenonycha dimorpha* in case someone else should inadvertently use the same name to describe a totally different species (a *homonym*) or accidentally describe the same beetle again as a completely new species under another name (a *synonym*).

As new information comes to light it is sometimes desirable to transfer a species from the genus in which it was originally described to another. When this occurs, the author's name is placed in parentheses. For example, the Ten-lined June Beetle was originally described as *Melolontha decemlineata* Say, 1823. Later, this scarab species was transferred to another genus and is now written as *Polyphylla decemlineata* (Say, 1823).

When describing a new species of beetle, researchers must designate one specimen as the *holotype* to serve as the name holder of the species. The holotype is the touchstone for all future researchers to verify the identity of a given beetle. Holotypes are placed in the care of museums or other public institutions where they are readily available to other researchers for examination.

The Filing System

With hundred of thousands of beetle species to describe and classify, it is essential that we have a universally recognized system that efficiently stores and facilitates the retrieval of data. The naming and arrangement of beetles into categories is called *taxonomy.* Taxonomists use categories known as *taxa* (singular *taxon*) and arrange them in a hierarchical system. The species taxon is the most exclusive of all animal taxa. Species are grouped into genera, which are arranged into tribes, which are organized into families, which are sorted into orders placed in classes. Classes are placed in *phyla* (singular *phylum*), which form the kingdom Animalia, the most inclusive animal taxon of them all.

Each of these taxa may be variously subdivided into smaller groups whose prefixes begin with *sub-* or *super-*. Some of these taxa have universal suffixes, such as *subtribes* (-ina), *tribes* (-ini), *subfamilies* (-inae), *families* (-idae), and *superfamilies* (-oidea).

Beetles are classified in the order Coleoptera. The elytra inspired the name of this taxon, derived from the Greek words *koleos,* meaning sheath, and *pteron,* meaning wing. The usage of the word Coleoptera can be traced back to Aristotle and was adopted by Linnaeus for use in his classification of insects. Because adult beetles possess three body regions, six legs, and external mouthparts, they are classified as insects. The class Insecta, including beetles, dragonflies, earwigs, true bugs, grasshoppers, termites, butterflies, ants, fleas, flies, and so on, is placed in the phylum Arthropoda. Arthropods also include arachnids, millipedes, centipedes, crustaceans, and horseshoe crabs. All are characterized by having exoskeletons and jointed appendages. By virtue of their multicellular bodies, their need to ingest complex organic matter, and their capacity to respond rapidly to events in their environment, beetles and other arthropods are classified in the kingdom Animalia (fig. 14).

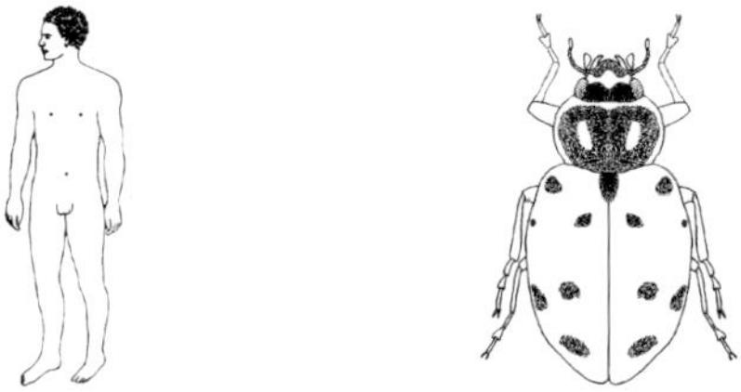

TAXON		
Kingdom	Animalia	Animalia
Phylum	Chordata	Arthropoda
Class	Mammalia	Insecta
Order	Primates	Coleoptera
Family	Hominidae	Coccinellidae
Genus	*Homo*	*Hippodamia*
Species	*Homo sapiens*	*Hippodamia convergens*

Figure 14. Classification of two familiar organisms: humans and the convergent ladybird beetle *(Hippodamia convergens)*.

Our schemes of classification, systems of nomenclature, and reliability of species identifications are not absolute. Earlier classifications will necessarily have to be revised as our understanding of taxa and their relationships improves. Species are combined, split, or moved to other genera as new information comes to light. Genera are combined, split, or transferred to other tribes and so on. This state of flux, so often frustrating to the general public who expect science to be immutable, demonstrates the flexibility and utility of the system as new data become available.

The Role of Systematics

Systematics is the study and ordering of natural diversity. It allows for identification and information retrieval. It also assembles taxonomic and other information for comparison, thus encouraging the blending of information from other biological disciplines. This matrix of data encourages new ideas and generates questions applicable to all fields of biological study. Because of continual philosophical and technological innovations such as cladistics and DNA sequencing, the field of systematics is a dynamic and exciting endeavor.

Information on beetle diversity provides more than just an understanding of their identity or evolution. Systematics also contributes information used in other studies of beetles, such as their *faunistics* (the study of some or all beetles in a region), *ecology* (the study of beetle interrelationships with their environment), and *zoogeography* (the study of beetle habitats and distributions, past and present).

Practically Speaking

Information generated by the various components of systematics has important application in the study of California beetles. Our knowledge of what a beetle is can sometimes provide clues to what it does. For example, all species of the longhorn beetle tribe Callidiini (Cerambycidae) feed exclusively on the redwood *(Sequoia sempervirens)*. Armed with these data it could be reliably assumed that any new species correctly classified in the same tribe would also feed on redwood. Systematics can therefore generate specific and intelligent theories by suggesting testable hypotheses about life cycles, habitat, and behavior deduced from the habits among species demonstrated to share common ancestry.

Predictions of species' properties based on the known habits of related taxa are incredibly important when choosing and implementing beetles, or their predators and pathogens, as biological control agents. Millions of dollars could be wasted on the introduction of inappropriate or ineffectual species.

Species distributions, such as those of ground beetles, provide clues as to the nature of past climates and the former spread of habitats, thanks to detailed faunistic and taxonomic studies. The fossil remains of beetles whose species or genera are still extant today, combined with information on their current habitat requirements, can be used with a considerable degree of confidence to infer ancient ecological conditions that prevailed tens or hundreds of thousands of years ago. The validity of these and other conclusions are all dependent upon the quality of current systematic knowledge.

Systematic data is also critical to our ability to manage and protect populations of beetles threatened with extinction. Laws and regulations drafted by legislators and other administrative bodies to protect sensitive species are more likely to succeed if they are based on accurate systematic information that accurately reflects the biology, distribution, and habitat requirements of the beetle.

CHAPTER 3

THE LIVES OF BEETLES

The amazing number of California beetle species is equally matched by the myriad ways in which they go about making their living. Each species has its own set of unique morphological, physiological, and behavioral adaptations wrought by the environment and, if they are lucky, that assure their reproductive success. Many engage in specialized and complex relationships with other organisms, either as consumers, predators, or parasites.

Because beetles make use of an astounding array of foods and habitats, they are both important consumers and recyclers of organic matter and thus play significant roles in the function of many ecosystems. With an understanding of the lives of beetles comes an appreciation of their ability to colonize and inhabit an incredible variety of California habitats.

Metamorphosis

Unlike most vertebrates that grow and develop with relatively little outward change in bodily form, beetles, like most other insects, undergo dramatic changes in body shape during the course of their lives. Like butterflies and moths, a beetle's life is marked by a series of successive changes called *complete metamorphosis,* a transformation marked by four distinct stages: egg, larva, pupa, and adult (fig. 15). Each of these stages allows beetles to exploit different aspects of their environment, giving them considerable flexibility in adjusting to ever-changing conditions. Parental care is rare among beetles, so adult and immature stages are seldom found together, and thus they do not compete with one another for food or space.

In the Beginning

Female beetles produce their eggs singly or by the hundreds. They scatter them about haphazardly or carefully deposit

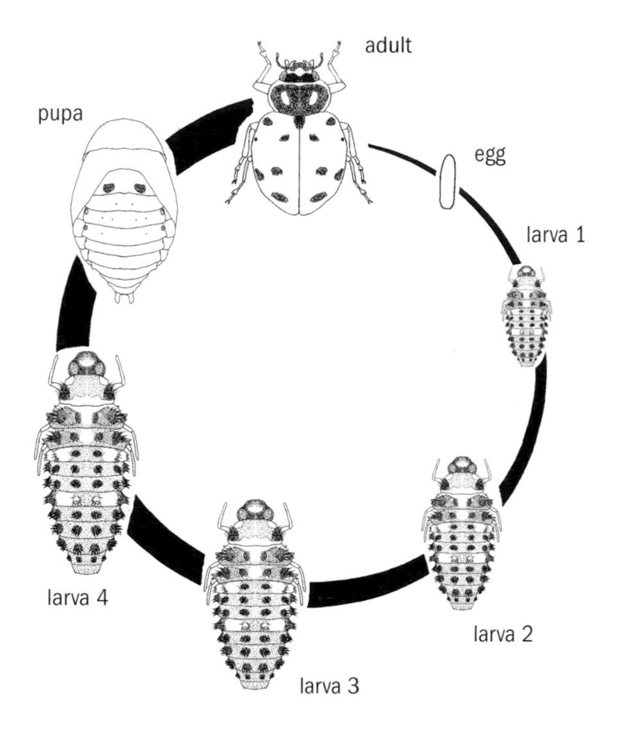

Figure 15. Life cycle of the convergent ladybird beetle *(Hippodamia convergens)*, illustrating complete metamorphosis.

each one with the aid of a long membranous tube, or *ovipositor.* Eggs are usually laid on or near suitable food for the hungry larvae. Many ground-dwelling scavengers lay their eggs in soil, dung piles, compost heaps, or other sites rich in decomposing organic material. Leaf grazers and miners drop their eggs at the base of the larval food plant, glue them to a stem or leaf, or carefully tuck them away within a crevice in the bark. Leaf-mining beetles, such as some metallic wood-boring beetles (Buprestidae) and weevils (Curculionidae) actually tear the leaf surface and sandwich their eggs in the narrow space between its surfaces. Some longhorn beetles (Cerambycidae) known as girdlers chew a ring around the terminal portion of a branch to create a source of dead wood in which their tunneling larvae develop. The freshly killed branch tips, particularly on oaks, are brown and stand in stark contrast to the surrounding green foliage, a phenomenon commonly referred to as *flagging*.

Sexless Eating Machines

The larvae of beetles begin their lives with just one purpose, to eat. Fueled by their ravenous appetites, most larvae grow rapidly and quickly outgrow their exoskeletons. The old exoskeleton must be regularly replaced with a newer, roomier model already forming inside their bodies. The stages between molts are called *instars*. There are usually at least three larval instars before reaching the pupal stage, but the larvae of some rain beetles (Pleocomidae) are known to molt seven to 13 times before pupating.

Each successive instar is generally similar in form, but larger in size. Blister beetles develop by a special type of metamorphosis called *hypermetamorphosis*, a process characterized by two or more distinct larval forms. The first instar, or *triungulin*, looks more like a silverfish than a typical beetle larva. It has long legs and is adapted for locating host insects. Those species that parasitize grasshopper egg masses burrow into the soil. Those that parasitize solitary bees climb to flowers and attach themselves to visiting bees to be transported back to the nest, where they feed on pollen and nectar. Upon finding the proper host, the triungulin molts into a less active larva with thick legs and begins to feed. The next molt produces a fat, C-shaped grub with short legs. After two or more molts the legless, nonfeeding larva rides out the winter. In spring another molt produces an active yet legless larva that spends most of its time preparing a pupal chamber.

Like their parents, beetle larvae consume a cornucopia of foodstuffs, including other insects, carrion, dung, and plants. Plant feeders consume either living or dead plant tissues, mining leaves or tunneling their way through roots, trunks, and branches. Although most California beetles produce one generation each year, some larvae that feed on nutritionally poor dead wood may take several years to complete their development. Furthermore, development may be delayed in response to dry or other stressful conditions. For example, the Black Polycaon Beetle (*Polycaon stouti,* Bostrichidae), which only lays its eggs on untreated wood, has been observed emerging after nearly 25 years from treated lumber used for furniture and house construction.

The Shape of Things to Come

Following the last larval instar, the beetle larva transforms into a pupa. The pupa serves as the vessel where dramatic biochemical transformations take place. During this stage, the tissues of a nonreproductive eating machine are transformed into a device equipped for mating and reproduction. In temperate climates the pupal stage, carefully tucked away in soil, in humus, or within the tissues of plants, is often better able to withstand freezing temperatures and thus serves as the stage during which some species pass the winter.

Despite the apparent universality of the distinct growth stages of complete metamorphosis among the Coleoptera, there are some notable exceptions among California's beetle fauna. *Paedogenesis,* or the retention of larval features in the reproductive adult, occurs in the phengodid beetles (Phengodidae) and pink glowworms (Lampyridae) (pl. 9). In these beetles the pupal stage of the female is barely distinct from the last larval instar. The adult female emerges from the pupa without wings and is distinguishable from the larva only by the presence of reproductive organs and compound eyes.

Plate 9. Adult females of the Pink Glowworm (*Microphotus angustus,* Lampyridae) (10 to 15 mm) retain larval features even when mature. These beetles are bioluminescent, emitting their light in a continuous glow. Photo by C. L. Hogue.

The Last Act

The final molt occurs when the requisite combination of temperature and moisture is achieved, triggering the emergence of the adult. Freshly emerged adults are said to be *teneral,* with soft, pale bodies still developing. Their wings, if present, are small and crumpled but soon expand and harden in a matter of minutes or hours. As the exoskeleton hardens, the body gradually takes on its normal color. Once they have reached the adult stage, beetles never molt again and are incapable of further growth, although the abdomens of some soft-bodied leaf and blister beetles are capable of some expansion as they become filled with food or eggs. Fully developed adult beetles are ready to mate and reproduce.

The Mating Game

Although some beetles, such as those infesting stored grain products, are literally wallowing in target-rich environments, others must disperse over wide distances and search through tangled vegetation or layers of leaf litter and soil to find a mate. With their relatively short adult lives coming to an end in a matter of weeks or months, beetles have little time to waste and have thus developed various channels of communication to maximize their efforts at finding a mate. They locate and seduce potential mates using sight, sound, or scent. Sometimes these strategies are remarkably effective, luring in numerous eager mates from considerable distances.

Lights On

A precious few of California's beetles are capable of *bioluminescence,* or producing their own light. *Adenosine triphosphate,* or ATP, powers the bioluminescence of pink glowworms and phengodids. An essential ingredient of all life forms, ATP is the universal currency of energy in living cells. Within the light-producing cells of glowworms and phengodids, the enzyme *luciferinase* attaches *luciferin* to ATP. This energizes the luciferin, enabling oxygen to attach to one if its carbon atoms, kicking out an electron into a higher orbit. The luciferin releases the oxygen and carbon together as carbon dioxide. As the electron drops back into its normal orbit, a bit of energy is released as a tiny flash of light. Multiply this activity by thousands of cells found within the light-producing organ of the glowworm and you have light that is easily visible to the human eye. The brightness and duration of the light is controlled by the amount of oxygen reaching the beetle's light-producing organs. Bioluminescence is nearly 100 percent efficient, with nearly all of the energy that goes into the system given off as light. In comparison, an incandescent lightbulb is only about 10 percent efficient, the remainder of the energy being lost as heat.

Scrapers and Headbangers

Several beetles produce sounds to attract mates and facilitate the act of mating. *Stridulation* is the act of rubbing one body part against another for the purpose of making a sound. Although the importance of sounds in courtship of California beetles is not well known, California longhorn beetles, June beetles (Scarabaeidae), and bark beetles (Curculionidae) stridulate by rubbing their elytra with their legs or abdomen to create chirping or squeaking sounds. They do this particularly when handled or attacked, so these sounds likely play a defensive role in some situations. Male death watch beetles (Anobiidae) bang their heads against the walls of their wooden galleries to lure females into their tunnels.

Blowing in the Wind

One of the most effective systems for bringing the sexes together is the use of pheromones. Plumes of messenger molecules are released and disperse into the air. Depending upon the species involved, pheromones are used by males or females to form mating aggregations that increase their chances of finding a mate. Male beetles that use pheromone systems to locate females, for example, the California Banded Glowworm (*Zarhipis integripennis,* Phengodidae), often have elaborate antennal structures packed with sensors capable of detecting just a few molecules. They fly in a zigzag pattern to follow the scent to its source. Calling females may be some distance away, tucked away among the tangled masses of vegetation or in a burrow in the soil.

The Mating Dance

Once a male and a female beetle have managed to locate each other, there is seldom more to be done. The male simply inserts his penis inside the female and ejaculates his sperm. Elaborate courtship behaviors are not common among beetles, although some species may engage in nibbling (e.g., soldier beetles [Cantharidae]) or antenna pulling (e.g., blister beetles [Meloidae]). The male generally mounts the female from above and behind.

A female beetle usually has an enormous reserve of eggs awaiting fertilization, but she generally needs to mate only once. Nevertheless, she is frequently courted and grappled by a host of enthusiastic males that respond to her pheromone plume. To make sure she receives enough sperm to fertilize her entire complement of eggs, she may store it in a sock-shaped reservoir in her body called the *spermatheca.* Fertilization does not occur until her eggs travel past the spermatheca, just before they are laid.

When the female has multiple partners, it is the sperm of the last male that fertilizes the eggs. To assure his paternity, the male may remain coupled with the female long after sufficient time has passed for the transference of his sperm. After ejaculation, the male tiger beetle (*Cicindela,* Carabidae) disengages his genitalia from his mate but continues to grasp her with his mandibles until she lays her eggs in the sand, a behavior called postinsemination association.

Not all species of beetles must mate to reproduce. *Parthenogenesis,* development from an unfertilized egg, occurs among several families of beetles, including leaf beetles and weevils. Males of parthenogenetic species are rare or unknown altogether. The females are solely responsible for maintaining the population simply by producing cloned offspring.

Doting Parents

Parental care among beetles is the exception rather than the rule. Dung scarabs (Scarabaeidae) and burying beetles of the genus *Nicrophorus* (Silphidae) exhibit varying degrees of parental care. Both males and females may cooperate in digging nests for their eggs and provision them with dung or carrion for their brood. Dung and carrion are rich in nutrients, and competition for these resources can be fierce. Many dung and carrion-feeding beetles have evolved tunneling or burying behaviors to quickly hide excrement or dead animals from the view of other marauding scavengers. Burial not only secures food for their young, it also helps to maintain optimum moisture levels for successful brood development.

Ambrosia beetles and bark beetles (Curculionidae) also provide food and shelter for their young, carving elaborate galleries beneath the bark of trees or in galleries that penetrate the sap wood. Adult beetles cultivate a fungus for food, much like some ants and termites. Females possess *mycangia,* specialized pits on their bodies where they store fungal spores as they fly about the forest in search of new trees to colonize. They introduce the fungus into the tree, later using it as food for both themselves and their larvae.

Feeding Behavior

Beetles are a major component of the so-called FBI—fungi, bacteria, and insects—the primary agents of decomposition. Equipped with powerful chewing mouthparts, both larval and adult beetles are capable of cutting, grinding, or boring through all kinds of plant and animal materials. Living or dead, there is a good chance that most plants and animals are eventually consumed by beetles. Their role in the physical and metabolic breakdown of organic matter is critical for recycling nutrients in almost every terrestrial ecosystem on the planet.

Herbivores

Many beetles are herbivores, getting nutrition by feeding on living plant tissue. Scarabs, leaf beetles, and weevils are particularly fond of leafy foliage and strip leaves of their tissues or completely defoliate plants. Pestiferous beetles are those species whose tastes include garden plants, shade trees, and agricultural or horticultural crops. Flowers are particularly attractive to some beetles, providing energy-rich pollen and nectar. Although not well studied, some California beetles may play an important role as pollinators.

Many wood-boring beetles feed on dead or dying wood as they tunnel. Dead tree branches of various sizes offer numerous feeding opportunities, from the bark inward. Beetle tunneling activities hasten the decay of the wood. As the level of decay increases, the riddled wood attracts a succession of species that prefer increasingly rotten wood. The combined tunneling activities of these beetles are essential in the recycling of nutrients.

Other beetles are scavengers, preferring instead to consume plant food cured by the action of fungi and bacteria. Dung scarabs prefer their plants partially broken down by the stomachs of horses, cattle, dogs, and other animals. Dung scarabs are poorly represented in California, but they are among the most beneficial and least appreciated of our insects. These industrious beetles not only consume animal feces, but in making provisions for their young, they also bury it.

Carnivores

Ground beetles and tiger beetles (Carabidae) are generally quick on their feet and often have powerful mandibles to overpower and tear apart insect prey. For their size, the larvae of tiger beetles are some of the most gruesome-looking creatures imaginable. They attack a broad range of insects and other invertebrates, including snails. Rove beetles (Staphylinidae) and clown beetles (Histeridae) are often found hunting for maggots and mites among leaf litter and carrion. Clown beetles are also found on trees, decaying plants, dung, sap flows, bird and mammal nests, and other habitats attractive to their prey. Lady beetles (Coccinellidae) consume a variety of foodstuffs, including pollen and molds, but are best known as predators of aphids, mealy bugs, and other plant pests. The larvae of phengodids prey on millipedes. Soldier beetles (Cantharidae) and checkered beetles (Cleridae) prey on plant-feeding and wood-boring insects, respectively. At least one bark-gnawing beetle (Trogossitidae) in California is an important predator of bark beetles and consequently has been introduced into other countries in an attempt to control infestations of lumber pests.

Some carnivorous beetles are specialists, preying only on certain kinds of insects. Some larval ground beetles and rove beetles actively seek out and consume the pupae of leaf beetles, whirligigs, and flies. Ant-loving scarabs (*Cremastocheilus* spp.) dine on the brood of ants.

Select groups of beetles attack and consume dead animals. Carrion beetles (Silphidae) scavenge freshly dead carcasses, occasionally preying upon fly maggots competing for the same juicy resource. Their cousins, the burying beetles, remove carcasses from the competitive fray by concealing the entire body underground. To hide beetles (Trogidae), the idea of a good meal is keratin-rich feathers, fur, claws, and hooves. Other species, such as ham and skin beetles (Cleridae and Dermestidae, respectively), prefer instead to gnaw on dried flesh. Natural history museums around the world use some species of skin beetles to clean animal skeletons for use in research collections and exhibits, whereas others are considered serious museum pests.

Symbiotic Relationships

Many beetles engage in symbiotic relationships with other organisms. *Symbiosis* refers to an intimate and specialized relationship between two or more organisms. A symbiotic relationship beneficial to both organisms is called a *mutualism.* Another form of symbiosis is *commensalism,* where one organism clearly benefits but the other is not adversely affected by the relationship. *Parasites,* on the other hand, live at the expense of their hosts.

Representatives of many beetle families, such as Staphylinidae, Histeridae, and Scarabaeidae, are found in and among the nests of ants and are known as *myrmecophiles,* or ant loving. Some myrmecophilous beetles are simply opportunists, living among the fringes of the colonies scavenging bits of food gathered by the ants. But others have managed through behavioral, chemical, or tactile mimicry to be accepted by their hosts and integrated into their social system.

Playing Together

Some beetles serve as hosts to symbiotic organisms living inside their bodies. *Endosymbiotic microorganisms* such as bacteria, yeasts, and fungi found in the guts of various wood-boring beetles are vital to their survival. These mutualistic microorganisms are essential for the digestion of the primary component of any plant food, cellulose. These symbiotic organisms generally reside in pockets found in the midgut. In stag beetles (Lucanidae) whose larvae develop in decaying wood, the fungal symbionts are found in a special fermentation chamber located in the hindgut. Wood-feeding larvae get their symbionts from their mothers, who cover their eggs in a residue laden with microorganisms. Upon hatching, the larvae immediately consume their own eggshells, ingesting the microorganisms.

Hitching a Ride

On the bodies of California's larger beetles are tiny, scorpion-like, yet tailless hitchhikers called pseudoscorpions. Like all arachnids, pseudoscorpions lack the ability to fly themselves. Instead, they enlist the aid of medium to large wood-boring beetles such as the Pine Sawyer (*Ergates spiculatus spiculatus,* Cerambycidae), the California Prionus (*Prionus californicus,* Cerambycidae), and June beetles of the genus *Polyphylla* (Scarabaeidae) to get from place to place. This type of commensalism is known as *phoresy.* Many pseudoscorpions develop beneath bark on tree trunks, hunting for small insect larvae and mites among the sawdust and galleries left by wood-boring insects. Once their hunting grounds are played out, the pseudoscorpions climb aboard another beetle, using it as transport to find newly fallen trees where food is abundant.

Tiny mites are also commonly found on some beetles. The phoretic mites found on burying beetles (Silphidae) feed on the eggs of carrion-feeding flies, reducing the competition for their hosts. Other species of mites have parasitic relationships with their beetle hosts.

Nibblers and Blood Suckers

The larvae of blister beetles, checkered beetles, and wedge-shaped beetles (Ripiphoridae) are all parasites of ground-dwelling insects. The larvae of all blister beetles and some checkered beetles attack the brood of ground-nesting bees or the egg pods of grasshoppers. Although the biology of wedge-shaped beetles is poorly known, the larvae are usually found in association with longhorn beetle larvae, solitary and social wasps, and cockroaches.

The most specialized and truly ectoparasitic beetle is the Beaver Parasite Beetle (*Platypsyllus castoris,* Leiodidae). Both larvae and adults live on beavers. These flattened, louselike beetles are perfectly adapted for spending their entire lives on beavers, where they feed on skin and body fluids.

Defense Strategies

Beetles are attacked and eaten by mammals, birds, reptiles, amphibians, fish, and other arthropods. The large size of some longhorn beetles, backed up by powerful mandibles, may be enough to deter some predators. But most beetles must rely on various behavioral and chemical methods of defense to avoid predation and infection. Feigning death is a behavioral strategy employed by some beetles. Others hide from their enemies or disguise themselves to look inedible. Some species resemble insects known to predators as difficult to capture or capable of inflicting painful stings, whereas others possess an arsenal of noxious chemicals to discourage predators.

Playing 'Possum

Desert ironclad beetles are so named because of their tough, thick exoskeletons that prevent dehydration and protect them from predators. The genus *Asbolus* (Tenebrionidae), nicknamed "blue death feigners," also use a defensive technique known as *thanatosis.* They simply stiffen their legs outward and play dead to avoid attracting further attention. Zopherid beetles (Zopheridae) also play dead when attacked (pl. 10). Some species even pull their legs and antennae up tightly into special grooves in their body. Faced with such an impenetrable fortress, small predators simply give up.

Hide-and-Seek

Beetles employ varying degrees of camouflage to blend in with their background and avoid detection by predators. Somber-colored brown or gray wood-boring beetles and

Plate 10. This zopherid beetle, *Zopherus tristus* (Zopheridae) (10.5 to 22 mm), plays dead when disturbed in an effort to avoid predation.

weevils blend in perfectly with the rough bark and gnarled branches of their food plants. Pale tiger beetles (*Cicindela*, Carabidae) almost disappear among the sandy shores of beaches, rivers, and streams. Even a few members of the usually brightly colored lady beetles are tan or striped, enabling them to remain undetected among the needles of pines. Bright, metallic green chafers of the genus *Dichelonyx* (Scarabaeidae), conspicuous anywhere else, meld seamlessly with needles and leaves.

Instead of blending in, some cryptic beetles attempt to look like something else altogether, possibly something inedible like feces. The larvae of tortoise beetles (Chrysomelidae) have a fork at the end of their abdomen in which to gather their own waste. When threatened, the larva flips this apparently distasteful accumulation over its back. Other tortoise beetles simply cover themselves in fecal pellets. The leaf beetles *Chlamisus* and *Exema* (Chrysomelidae) combine their irregular body surfaces and chunky physique to resemble the fecal pellets of caterpillars.

Boo!

Bold designs such as eyespots or sudden flashes of bright colors are thought to startle or confuse would-be predators. The outsized eye spots of the Eyed Click Beetle (*Alaus melanops*, Elateridae) may momentarily confuse a predator in terms of how to direct a sneak attack, allowing the beetle an extra moment or two to escape. Many dull-colored metallic wood-boring beetles (Buprestidae) reveal flashes of bright iridescent blue, green, or red on their abdomen as they lift their elytra to take flight. The abdomens of tiger beetles are similarly colored. The sudden appearance of these bright colors is thought to startle some predators.

Mimicry

California's beetle fauna has several species that mimic the appearance or behavior of stinging or distasteful insects. The flightless, black Cactus Longhorn Beetle *(Moneilema semipunctatum)* resembles some foul-smelling and foul-tasting darkling beetles (Tenebrionidae) both in appearance and behavior. When attacked, many darkling beetles raise the tip of their abdomen into the air to release a foul-smelling defensive fluid. When similarly attacked, the Cactus Longhorn Beetle also stands on its head but lacks any noxious chemicals to back up the implied threat. For added protection, this beetle spends much of its time surrounded by the spines of its food plant, the cholla *(Opuntia)*.

Some longhorn beetles, especially those that visit flowers, are striking mimics of bees and wasps. Their bright contrasting colors are often patterned, giving them the narrow-waisted appearance of stinging insects. The Lion Beetle

Plate 11. The Lion Beetle (*Ulochaetes leoninus,* Cerambycidae) (17 to 32 mm) looks and behaves just like a bumblebee, even attempting to sting with its ovipositor. When disturbed, the Lion Beetle raises its abdomen forward over its back while flapping its wings, reinforcing its bee-like appearance.

(*Ulochaetes leoninus,* Cerambycidae) (pl. 11) looks and behaves just like a bumblebee (*Bombus* spp.), even attempting to sting with its ovipositor. The bumblebee scarabs *Lichnanthe* (Glaphyridae) also mimic bees. These hairy beetles buzz fast and low over the ground, resembling a bee in flight. Several species of checkered beetles are boldly colored to resemble pugnacious ants or wingless wasps known as velvet ants. Their quick, jerky movements further reinforce the charade.

Blistering and Burning Defenses

Beetles generally advertise the fact that they contain noxious chemicals, either by bold warning patterns known as *aposematic coloration,* or with distinctive chemical release behaviors, such as standing on their head. These chemical arsenals are widespread in beetles and are used as repellents, insecticides, or fungicides and are directed against a wide range of target organisms. These chemical defenses are usually produced by glands in the body and stored in special chambers or within the blood. Still other species sequester the chemical defenses of their food plant and incorporate them into their own defense system.

Under attack, darkling beetles of the genus *Eleodes* stand on their head as a defensive gesture before expelling noxious-smelling quinones from their anus, but some rodents are seldom deterred by the smelly defenses. They simply grab the insect with their paws, force the distasteful tip of the abdomen into the soil and, beginning with the head, enthusiastically consume its tastier tissues.

Many ground beetles possess defensive glands just inside the tip of their abdomen that produce a variety of hydrocarbons, aldehydes, phenols, quinones, esters, and acids. These compounds are released from the anal opening as a noxious stream of fluid. Bombardier beetles (*Brachinus* spp.) store the components of their chemical arsenal in separate chambers. When attacked, the hydroquinones, hydrogen peroxide, peroxidases, and catalases are injected into a third chamber. The resultant synergistic reaction explodes out of the body with an audible pop, producing a small yet potent cloud of acrid spray that is literally at the boiling point. With the aid of an incredible anal turret, this toxic cloud is directed very accurately at the bombardier's enemies.

Blister beetles produce a caustic defensive agent known as cantharidin. Cantharidin is stored in the blood and is secreted

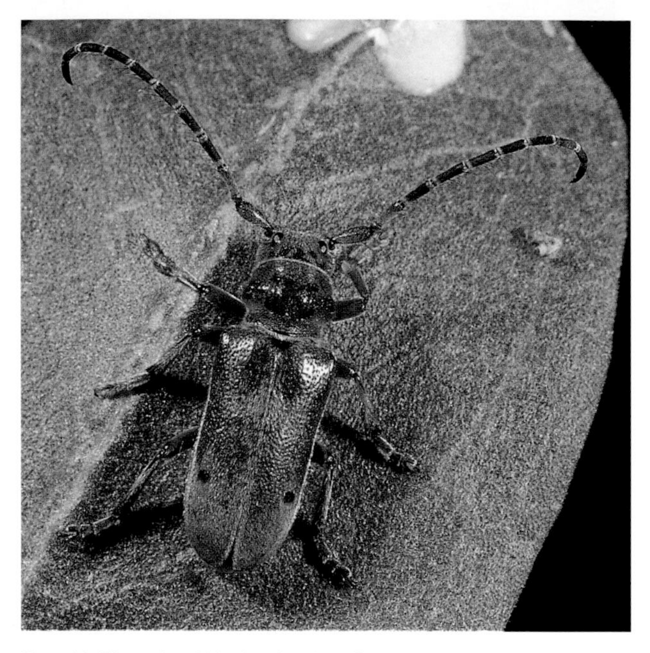

Plate 12. The red and black coloration of this milkweed borer, *Tetraopes femoratus* (Cerambycidae) (8 to 19 mm), advertises its distastefulness to potential predators.

from membranes around the base of the legs, a phenomenon known as reflex bleeding. Cantharidin causes blistering and sores on the softer internal and external tissues of vertebrates. Antlike flower beetles of the genus *Notoxus* (Anthicidae) actively seek out their cantharidin-bearing cousins to chew on their elytra and other body parts to acquire their own chemical defense system. They are often seen swarming over the bodies of dead or dying blister beetles.

Other beetles also acquire their defensive chemicals from outside sources. The bright red and black-spotted milkweed borers (*Tetraopes* spp., Cerambycidae) (pl. 12) sequester toxic cardenolides (cardioactive steroids) from the milky sap of milkweeds. By shunting the harmful compounds out of their digestive tract and into their body wall, these beetles become distasteful to most predators. The large and beautiful Blue Milkweed Beetle (*Chrysochus colbaltinus*, Chrysomelidae) apparently sequesters very little if any toxins found in milkweed tissue and sap. The protective coating of feces applied to the eggs, however, is quite concentrated with the chemical.

CHAPTER 4
DISTRIBUTION OF CALIFORNIA BEETLES

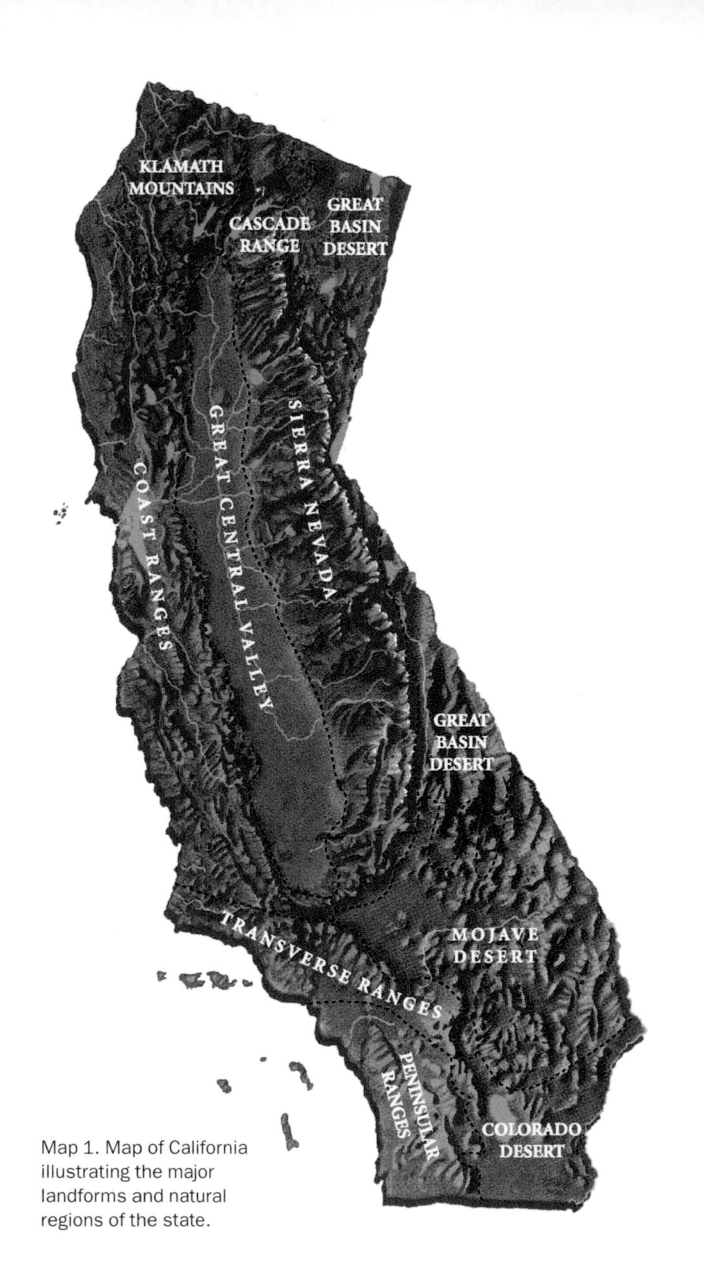

Map 1. Map of California illustrating the major landforms and natural regions of the state.

The distribution of California's beetles and how they came to be where they are today is complex and only partially understood. Over millions of years they have settled into the state's mountains, deserts, valleys, and islands as a result of nearly imperceptible adaptations to an ever-changing environment. Their ancestors, who originated from three distinct regional faunas, had to cope simultaneously with the gradual appearance of formidable mountain ranges, vanishing inland seas, and changing climate.

Before we can begin to appreciate the distribution of beetles in California, it is important to understand the incredibly diverse lay of the land. Even the most casual traveler can see that the state is divided into four basic zones: mountains, the Great Central Valley, deserts, and islands (map 1). Each of these physical features serves as a corridor of dispersal for some beetles and acts as a barrier to others. Their unique and shared vegetation communities, soil types, and climates act in concert to create seemingly infinite habitats that support a breathtaking diversity of beetles.

Mountains

The mountains of California are part of an extensive system of highlands covering much of western North America. Beginning in extreme northwest Alaska and extending southward to the Isthmus of Panama, this system is one of the longest mountain chains on Earth. With their axes generally running north and south, the mountain ranges create formidable barriers to the east-west movements of beetles yet simultaneously provide a thoroughfare for north-south dispersal. The mountains of California are divided into five regions: the Sierra Nevada, the Pacific Northwest, and the Coast, Transverse, and Peninsular Ranges. Of these, only the Transverse Ranges, which include the Santa Monica, San Gabriel, and San Bernardino Mountains, run east to west.

Their respective locations, exposure to sunlight, amount of precipitation, soils, and vegetation are distinctive and determine the nature of their beetle faunas.

Great Central Valley

The Great Central Valley is composed of the Sacramento Valley to the north and the San Joaquin Valley to the south. They combine to form a single north-south valley separating the Coast Ranges from the Sierra Nevada. More than any other region of the state, the Great Central Valley has been greatly altered by human activity. Nearly all of the original grasslands, or prairie, is gone. It has been replaced by farmland and introduced grasses. Most of the low-lying marshes are also gone, as well as the gallery forest lining the major rivers and their tributaries.

Like the deserts, the Great Central Valley is one of California's driest regions. Positioned in the rain shadow of the Coast Ranges, the winter rainfall ranges from moderate in the north to light in the south. Much of the low-lying areas in the southern portion of the valley support desertlike communities and share several species of beetles that occur in the deserts, especially darkling beetles (Tenebrionidae) and weevils (Curculionidae).

Deserts

California has three deserts: the Great Basin, Mojave, and Colorado. All were formed by the rain shadows of mountain chains lying to their west or south. Receiving less than 250 mm (10 in.) of rain annually, each of these deserts experience extreme temperatures. High winds and dry, clear air enhance the intensity of light reaching the ground, increasing evaporation rates. Nutrient-poor, alkaline soils, often a by-product of dry climates, are sparsely vegetated. California's deserts vary in terms of temperature extremes and the timing of precipitation.

California Islands

Sparsely populated by humans, California's islands serve as refuges for species once broadly distributed but now absent, or nearly so, from the mainland. As with the mainland, grazing, agriculture, and introduced species have caused significant alteration and damage to their native island floras and faunas. These islands have long fascinated naturalists and are considered evolutionary laboratories for biologists studying the ecology and distribution of plants and animals. All of California's islands are now in the hands of public agencies, organizations, and conservancies dedicated to returning them to their original conditions.

Because of their relative inaccessibility, California's Channel, Farrallon, Año Nuevo, and the San Francisco Bay Islands have been poorly sampled by entomologists. Their known fauna are closely related to that of the adjacent mainland, a feature typical of continental islands. Most of the entomological work conducted on the islands has focused on butterflies, moths, and grasshoppers.

The first list of beetles known from the California Channel Islands appeared in 1897 and was compiled from literature records and incidental collections made by biologists studying plants and birds. The first detailed insect survey of the islands was conducted from 1939 to 1941 by the Natural History Museum of Los Angeles County. The second flurry of entomological activity on the islands occurred in 1966 with the establishment of a University of California field station on Santa Cruz Island. Since then, numerous institutions throughout the United States have sponsored collecting trips to the islands. In spite of these efforts, the beetles of the islands remain poorly known. Of the more than 300 species of beetles recorded from the islands thus far, about 40 are thought to be endemic, occurring nowhere else.

Many flying beetles probably reached the islands by air, blown out across the channel by strong winds. The shorter distances between islands and the mainland associated with lower sea levels during the Pleistocene (40,000 to 10,000 years ago) increased the chances of successful beetle introductions from the mainland. Flightless beetles probably drifted out to the islands on floating debris. During flood years, swollen rivers can carry entire trees and their inhabitants out to sea. The Santa Clara, Los Angeles, San Gabriel, and Santa Ana River systems flush massive amounts of debris downstream and out into the Pacific. Light bodied and naturally buoyant, beetles and other insects stand a much better chance of becoming established on offshore islands than larger, heavier animals such as reptiles and mammals.

Patterns of Distribution

While collecting beetles one soon recognizes that certain species, especially those found high in the mountains, are related to others living in similar habitats elsewhere in the Northern Hemisphere. By studying the relationships and distributions of beetles, researchers have discovered that California's beetle diversity is derived from at least three basic regional faunas. The Vancouveran fauna is northern in origin and links California's beetles with those found in the Pacific Northwest, Rocky Mountains, Canada, and Eurasia. The southern Sonoran fauna connects the state's beetle fauna with Arizona and northwestern Mexico. The California fauna is represented by species that apparently evolved in California, independently of the northern and southern faunal elements and in isolation from the rest of western North America.

During the cooler and wetter times of the past, elements of the Vancouveran fauna of the north extended their ranges southward into central and southern California. As the climate became warmer and drier over the past 10,000 years, these northern species that preferred cooler, wetter habitats either retreated northward or managed to find refuge in isolated pockets of suitable habitat. During these same periods, Californian and, to a lesser extent, Sonoran faunal elements followed the Vancouveran contraction, expanding their ranges northward. The repeated expansion and contraction of beetle ranges, driven by ever-changing climatic conditions, resulted in a patchwork of refugia for all three faunal groups throughout the state.

Northern Faunal Influences

Many beetle genera of the Vancouveran beetle fauna commonly found in California also occur in the mountains of Europe, northern Asia, the Rocky Mountains, and to a lesser extent, eastern North America. These regions drew their assemblages of beetles from a single great fauna that spread throughout much of the Northern Hemisphere. For example, ground beetles of the genus *Calosoma* (Carabidae) are found in Asia. The predaceous diving beetle genus *Dytiscus* (Dytiscidae) includes species ranging from Arctic habitats to the northern slopes of the Himalayas and North Africa in the Old World and southward to Guatemala in the New World. The genus *Agabus,* another predaceous diving beetle, is also found in Europe. The Rugose Stag Beetle (*Sinodendron rugosum,* Lucanidae) has relatives in Europe and the Middle East. Click beetles such as *Elater* and *Limonius* are found in Europe. The longhorn beetle genera *Ergates* and *Rosalia* (Cerambycidae) both have species in Europe and Japan. The longhorn genus *Spondylis* consists of three species, occurring in Vancouveran California, Eurasia, and in the highlands of central Mexico. Some species are distributed in both Europe and North America, including California, such as the longhorn beetles the Hairy Pine Borer *(Tragosoma desparius)* and Ribbed Pine Borer *(Rhagium inquisitor)*, both associated with conifers.

During the Pleistocene, the expansion and contraction of glacial ice broke up this northern assemblage of species. Advancing sheets of ice forced many beetle species southward along the Pacific Coast and western mountains of North America. As conditions warmed, some species withdrew northward with the retreating ice, whereas others took refuge in scattered pockets of suitable habitat that persisted in the higher elevations of southern mountain ranges and along the northern Pacific Coast. This is how many northern Vancouveran beetles reached their southernmost ranges in the Coast, Transverse, and Peninsular Ranges, including snail eaters *(Scaphinotus)*, stag beetles *(Sinodendron)*, rain beetles (*Pleocoma*, Pleocomidae), metallic wood-boring beetles (*Chalcophora*, Buprestidae), and longhorn beetles (*Centrodera*, *Ulochaetes*, and *Ergates*).

Today the Vancouveran faunal region includes the Aleutians, southern Alaska, coastal British Columbia, and western Washington and Oregon. In California the Vancouveran region includes primarily northwestern California and western slopes of the Cascade Range and Sierra Nevada, as well as portions of the Coast, Transverse, and Peninsular Ranges.

The humid coastal Vancouveran, or Pacific maritime subregion, includes the Klamath Mountains and the coastal strip running southward from the Oregon border to the middle of Monterey County. Many of the ground beetles, click beetles, fireflies, stag beetles, and weevils found here also occur in the cool, wet forests west of the Coast and Olympic Ranges of Oregon, Washington, and British Columbia. Although many longhorn beetles, such as the Pine Sawyer *(Ergates spiculatus*

spiculatus), occur in both the humid coastal and arid Sierra Nevada–Cascade Range subregions, there are numerous examples of species that occur in one or the other region, but not in both.

The arid Sierra Nevada–Cascade Range Vancouveran subregion runs along the western flanks of the southern Cascade Range and Sierra Nevada. It is also found in isolated patches high in the Transverse and Peninsular Ranges, ending in the San Pedro Martir of Baja California, Mexico. Ground beetles, predaceous diving beetles, rove beetles, and click beetles are especially rich in this region and are typically associated with coniferous forests dominated by Jeffrey pine *(Pinus jeffreyi)* and ponderosa pine *(P. ponderosa)*. At elevations from 2,100 m (7,000 ft) in the Cascade Range to nearly 3,000 m (10,000 ft) in the southern Sierra Nevada are species of ground beetles with distributions in Alaska, Canada, the Rockies, and across North America to the northeastern United States. Widely separated populations of beetles or closely related species are found both high in the Sierra Nevada and in similar habitats in the Transverse and Peninsular Ranges.

Southern Faunal Influences

California beetles of Sonoran origin are for the most part found in the deserts and are variously adapted to living under extreme conditions of temperature and drought. The Inflated Blister Beetle (*Cysteodemus armatus*, Meloidae), darkling beetles (*Asbolus, Eleodes, Eusattus, Stenomorpha*, Tenebrionidae), and some weevils (*Ophryastes, Scyphophorus*, Curculionidae) are all wingless to conserve water and insulate their bodies from high temperatures.

The nearest relatives of California's Sonoran fauna live in the hot, dry uplands of Arizona, New Mexico, parts of Texas, and northern Mexico. The longhorn beetle *(Archodontes melanopus aridus)* also occurs in Arizona, with two additional subspecies ranging across the southern United States and southward into Mexico: *A. m. melanopus*, which ranges from Florida to Texas, and *A. m. serrulatus*, which ranges from Texas to Arizona. Another longhorn beetle, the Mesquite Borer *(Derobrachus geminatus geminatus)*, is found throughout the Mojave and Colorado Deserts and occurs eastward into Arizona, New Mexico, and Texas.

Some members of the Sonoran beetle fauna share affinities with the neotropical fauna typical of more southern climates. For example, the nearest relatives of Sleeper's Elephant Beetle *(Megasoma sleeperi)*, of the Colorado Desert, occur in Arizona, Texas, and Baja California Sur. Other species of *Megasoma*, some of which are among the largest and heaviest beetles in the world, are found throughout mainland Mexico southward into Central and South America. A nocturnal tiger beetle, *Tetracha carolina carolina* (Carabidae), is distributed across the southern half of the United States, Mexico, and the West Indies, whereas its nearest relatives range southward into South America. The genus *Zopherus* (Zopheridae) with four species in California, ranges across southern United States southward into Central America. Interestingly, one species, *Z. sanctahelenae*, is known only from Napa County, possibly marooned there by receding Sonoran habitat.

Elements of the Sonoran fauna extend from the Mojave Desert through the Walker Basin of Kern County and into the southern San Joaquin Valley, particularly on the west side. This desert island is home to species of darkling beetles (*Edrotes* and *Eleodes*) and weevils (*Ophryastes* and *Eupagoderes*) that are typical members of the Sonoran fauna.

A derivative of the Sonoran fauna, the Great Basin fauna extends throughout the entire lowland area between the Rocky Mountains and the Sierra Nevada. The Cactus Longhorn Beetle *(Moneilema semipunctatum)* is found throughout the Colorado, Mojave, and Great Basin Deserts, as well as in desert extensions reaching into the coastal plain of southern California. Other members of the genus are found in parts of the midwest, the southwest, and into Mexico.

California Faunal Influences

The last important faunal assemblage in the state was not greatly influenced by the northern Vancouveran fauna or the southern Sonoran fauna. Instead, this group of beetles apparently evolved in California in isolation from the rest of North America. The Californian beetle fauna has been long established in much of the central and southern parts of the state, especially on coastal plains of southern California, the southern Coast Ranges, and the Great Central Valley.

The small, wingless longhorn beetle *Ipochus* is distinctive among the North American longhorn beetle fauna and is known from only two species along the Pacific Coast of California and Baja California Norte. *Ipochus fasciatus* is found in the San Francisco Bay Region, the surrounding foothills of the southern California coastal plain, and on several of the Channel Islands. Two species of the zopherid genus *Phloeodes* are primarily associated with oak woodlands of California.

The scarab beetle genera *Coenonycha* and *Paracotalpa* are represented by species occupying both montane and desert habitats of Arizona, California, Idaho, Nevada, Washington, Utah, and Baja California. They reach their zenith in terms of species diversity in the southern Coast Ranges, San Joaquin Valley, and chaparral regions of southern California.

Some elements of the California fauna are restricted to coastal dunes and their remnants. Weevils of the genus *Trigonoscuta* occur along the coast of Washington, Oregon, and California. They also occur along the ancient beach line of the long-defunct Lake Cahuilla, visible along the base of the Santa Rosa Mountains west of the present-day Salton Sea. Darkling beetles of the genus *Coelus* live only among sand dunes and range from British Columbia to northern Baja California peninsula. Like *Trigonoscuta*, some populations of the tenebrionid, or darkling beetle, *Coelus* (e.g., *Coelus ciliatus*) are restricted to inland dunes once part of a coastal system. *Coelus* also occurs on inland dunes on the western side of the San

Joaquin Valley in eastern Monterey County. These dunes are thought to be fragmented remnants of sandy beaches that bordered the Pacific Ocean when it occupied the Great Central Valley during the late Pliocene or early Pleistocene.

The Future of California Beetle Biogeography

In spite of nearly 200 years of study and collecting in California, researchers have only begun to scratch the surface in terms of fully understanding where beetles live and how they got there. With vast tracts of California remaining terra incognita for beetles, especially its mountain tops, eastern deserts, and islands, much work needs to be done. Furthermore, improved trapping and sampling techniques conducted year-round will certainly reveal new or rarely encountered species in areas previously surveyed.

Additional beetle species may be added to the California state list at any time, either by the discovery of species new to science or by natural range extensions of populations currently residing in adjoining states or Mexico. Agricultural development in the Coachella Valley led to the short-lived westward expansion of a cotton pest from Arizona, the Boll Weevil *(Anthonomus grandis)*. Another recent immigrant from Arizona, the scavenger scarab beetle *(Hybosorus illigeri)*, was recently collected in the Colorado Desert. A native of southern Europe, this species was apparently introduced into the United States around 1848, quickly established itself throughout the southeastern United States, and has been moving westward ever since.

Other exotic species from outside the United States have been deliberately introduced into California, several of which are now established locally or have become widespread. The introductions of leaf beetles (Chrysomelidae) and ladybirds

(Coccinellidae) as biocontrol agents are encouraged in an effort to control pest plants and insects (see chapters 5 and 6). This practice is designed to wean our society from dependence on costly and potentially harmful herbicides and insecticides. But the introduction of exotic species as biological control agents not only affects target pest species, it may also have unforeseen consequences on native species as well. Self-propagating and often unpredictable in their dispersal, some introduced beetles can easily displace native species unable to compete for food or egg-laying sites (see chapter 5).

With more than 1,600 km (1,000 mi) of exposed coastline and heavy international traffic through its harbors and airports, it is no wonder that California is fertile ground for beetle species immigrating from other parts of the world. Many of the state's exotic beetles originated in habitats from eastern Asia and parts of Europe with remarkably similar climates. After establishing populations around major harbors and airports in San Francisco, Los Angeles, and San Diego, some species have become well established and continue to expand their ranges throughout the state.

Most accidentally introduced species do not become established, but after repeated introductions under favorable conditions, it is only a matter of time before a new and potentially pestiferous exotic beetle becomes established in California. This is why local, state, and federal agencies are charged with regulating the movement of materials passing through California ports and regularly trap and inspect shipments for beetles and other insect pests.

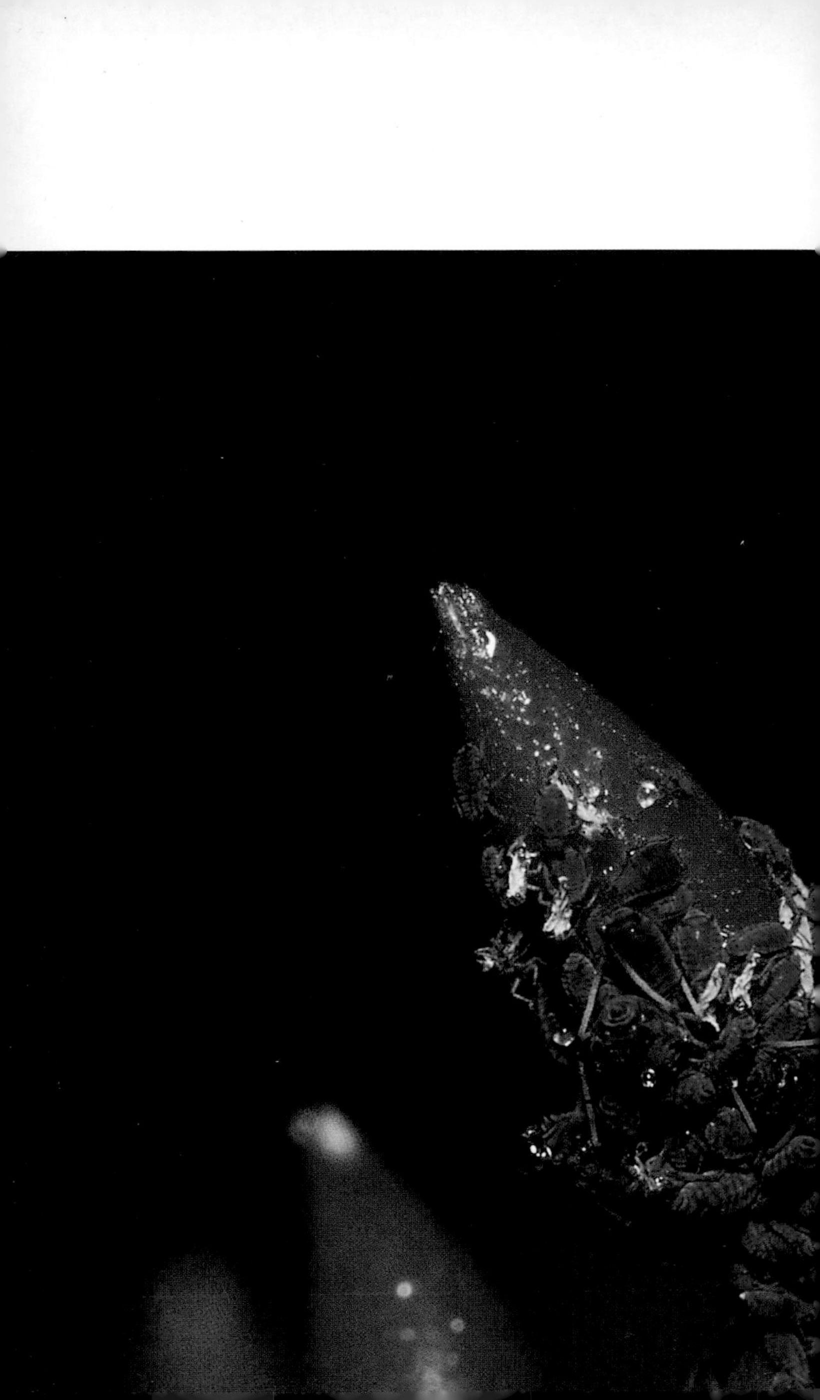

CHAPTER 5

BEETLES OF SPECIAL INTEREST

Some of California's beetles capture the attention of researchers, naturalists, and homeowners not because of what they look like, but because of what they do. For example, the remains of beetles thousands of years old can reveal particular details of ancient environments far better than the bones of large, extinct mammals. Highly specialized species with limited distributions, some of which are under federal protection as threatened or endangered species, are red flags for entire habitats that are in need of protection. Indigenous forest beetles, breaking down wood into its basic components, are often part of a delicately balanced ecosystem. When their recycling activities are directed at trees in managed forests, however, these beetles can become serious pests, damaging or destroying millions of dollars of lumber annually.

Other beetle pests are sure to attract attention, especially those species originating from other parts of the world. To these species California must, at first glance, resemble an impenetrable island fortress. Surrounded by high mountains, dry deserts, and a vast expanse of ocean, only the most persistent of beetles could ever gain a tarsal hold in the state. Nearly all of California's exotic beetle pests arrived with the help of humans, however, accidentally or otherwise. More than two centuries ago tiny scavengers arrived in California with goods packed on European ships and have been invading homes, cupboards, and pantries in search of dried goods ever since.

Pests from as far away as Australia and Asia have found their way to the Golden State, sometimes via rather circuitous routes. Some of these exotic beetles readily settled in urban landscapes filled with similarly exotic plants, whereas aggressive detection and quarantine programs thwarted the establishment of others. But not all beetle immigrants are considered pests, at least, not yet. Many of California's dairy farmers, equestrians, and dog owners apparently benefit from the activities of African and European species purposely introduced to bury animal waste.

Fossil Beetles

California's fossil beetles provide researchers with sensitive tools capable of revealing complex environmental changes in the Earth's not-so-distant past. Their bodies are preserved in sticky asphalt, prehistoric mammal nests, and mineralized nodules. Seldom found intact, their broken and scattered remains often possess an amazing degree of detail sufficient for their placement to species. By carefully comparing the fossil remains of beetles to modern species, scientists are able to infer the nature of past ecological conditions with a considerable degree of confidence. The abundance of preserved beetles combined with the availability of data on modern species give beetles a fundamental role in unlocking the mysteries of the Pleistocene extinction that befell so many mammals some 10,000 years ago.

Entombed by Asphalt

California's asphalt deposits are perhaps best known for their fossil vertebrate remains. Beginning in 1905, amateur fossil hunters and early paleontologists dug more than 100 pits and carted away tons of bones just west of downtown Los Angeles in an area that would become known as Rancho La Brea, or the La Brea Tar Pits. Impregnated and preserved by asphalt, these skeletal remains belonged to various late Pleistocene mammals that roamed southern California some 10,000 to 40,000 years ago. More than 80 species of insects have also been recorded from the site, many of which are beetles.

The name La Brea "Tar Pits" is somewhat of a misnomer because the large, bubbling pools visible today are not natural features but are instead the abandoned excavations of early fossil hunters. Left open, these excavations were eventually filled by intermittent springs of water and asphalt. The popular notion that these and other asphalt pools lured hapless animals to a slow and sticky death, which in turn attracted predators and scavengers that also became mired in the goo, has given way to new evidence. The scuffed and worn condition of many bones indicates that they were washed down by torrential rains and accumulated in floodwater deposits. The bones were then saturated and preserved by asphalt forced through the surrounding sediments by methane gas.

The first descriptions of beetles preserved in the asphalt appeared in 1908, but it was not until the 1940s that scientists began to focus their attentions on the remains of plants, insects, and other small animals as a means of studying prehistoric climates. Here their remains have not been replaced with minerals, in contrast to the much older bones of dinosaurs or the wood of a petrified forest.

The proper collection of microfossils, including those of insects, at Rancho La Brea is exacting and time-consuming work. The excavation is laid out in meter-square grids within a coordinate system so that the location and position of each bit of organic material and its surrounding substrate are carefully recorded. Once separated from the surrounding matrix, the fossil remains of beetles are washed with solvents in ultrasonic cleaners. Each fragment is then assigned a catalog number and stored dry in a clear gelatin capsule, such as those used for drugs and vitamins. The capsules are in turn stored in glass vials.

Determining the age of fossil beetles with radiocarbon dating techniques is possible only by indirect means. The age of beetle fragments is estimated based on radiocarbon dating of bone and wood samples found in the associated matrix. Because beetles and other small animals are still being trapped today, care must be taken when selecting samples to date because gases percolating through the asphalt can mix modern bits with the Pleistocene matrix.

The presence or absence of beetles in fossil deposits might represent real changes in faunal composition or simply represent a bias in attraction to the site or subsequent preservation.

Ground beetles (Carabidae) and darkling beetles (Tenebrionidae) are common in the deposit. The remains of predaceous diving beetles (Dytiscidae) such as the Giant Green Water Beetle *(Dystiscus marginicollis)* and giant water scavengers (Hydrophilidae) such as the Giant Black Water Beetle *(Hydrophilus triangularis)* are also frequently found in the deposits. They were apparently drawn to the shiny surfaces produced by oily asphalt seeps or accumulations of water on their surface. Some beetles may have been attracted to the asphalt itself. Scavengers, such as rove beetles (Staphylinidae) and carrion beetles (Silphidae), probably became mired in the asphalt because they were attracted to carrion or other organic materials already trapped in the seeps.

Early workers described the beetle remains as extinct species, but subsequent studies have shown that nearly all specimens represent modern species. Species thought to be extinct include *Copris pristinus* and *Onthophagus everestae* (Scarabaeidae), both dung scarabs whose nearest living relatives are found today in Arizona and Mexico. These dung beetles apparently vanished from California along with the disappearance of the producers of their food source, the large mammals. Like some of the other local plants and animals, they were also probably adversely affected by the overall drying of the climate and subsequently disappeared. These beetles may be found alive somewhere in the less explored regions of Mexico. Another fossil dung scarab that no longer occurs in California is *Phanaeus labreae.* It cannot be assigned to a modern species because of the poor condition of the only known specimen, a highly distorted head capsule. The chafer *Serica kanakoffi* (Scarabaeidae) was also described from a head capsule, a structure inadequate for use in species placement. The fossil darkling beetle *Coniontis remnans* (Tenebrionidae) has not been recognized among the modern fauna, but this does not mean that it is extinct. It belongs to a poorly known genus with several undescribed species, one of which may turn out to be the same species as the fossil.

Other California asphalt deposits have produced fossil beetles. The Carpenteria deposit, located on a coastal bluff in Santa Barbara County, was abandoned as an asphalt quarry when fossils were first discovered there in 1927. Efforts to quarry the site were eventually abandoned, and the site was used as a refuse dump beginning in the 1940s, obscuring the precise location of the early fossil finds. The remains of beetles extracted from the deposits near Maricopa and McKittrick, both located in the southern San Joaquin Valley of Kern County, have also been studied.

Time Capsules

Spider beetles (Anobiidae) are scavengers of plant and animal materials typically found in the nests of insects, birds, and mammals, especially rodents. They are also known to feed on the dung of rodents and may consume bits of their hair, skin, and other parts. A fossil spider beetle, *Ptinus priminidi*, was described from specimens found in fossil wood rat *(Neotoma)* nests in the dry mountains of San Bernardino County in southeastern California.

Wood rats are common today in some parts of the state. Their nests, consisting of accumulations of organic debris, are referred to as *middens.* Middens are constructed primarily of plant materials gathered from the immediate vicinity of the nest, including cactus pads and the occasional animal bone. Middens often form a representative cross section of the more conspicuous plant species in the region. The accumulation of plant materials attracts other animals seeking food and shelter, especially insects. Over time the remains of generations of the midden's inhabitants accumulate, sealed by rat urine and preserved for thousands of years among protective rock crevices.

Radiocarbon dating of the plant materials found in middens suggests that *Ptinus priminidi* lived between 12,300 and 30,000 years ago. The specimens are all remarkably well preserved. One notable specimen is coated in a protective glaze of petrified rat urine, much like a Siberian mammoth preserved in permafrost. Whether this species is extinct or still alive and kicking in the nests of modern wood rats somewhere in western North America is not yet known.

Stoned Beetles

Fossil insects are also encased inside mineral deposits. Calcareous nodules formed in association with algae on ancient lakebeds tainted by volcanic waters and gases have been found on Frazier Mountain (Ventura County) and in the Tehachapi (Kern County) and Calico Mountains (San Bernardino County). A single larva of a skin beetle (Dermestidae), *Miocryptorhopalum kirkbyae,* was extracted with formic acid from a nodule collected from Upper Switchback Canyon in the Calico Mountains.

Sensitive Species

Undoubtedly, the activities of humans seriously impact California's flora and fauna, including beetles. Habitat loss as a result of urban development, grazing, agriculture, and the introduction of exotic species all contribute to the local disappearance, or *extirpation,* and extinction of the state's beetles. California is particularly susceptible to species loss because of the sheer number of beetles specifically adapted to living in the state's abundance of unique habitats. At least three species or subspecies of California beetles are believed to have become extinct in the last 150 years: the Oblivious Tiger Beetle (*Cicindela latesignata obliviosa,* Carabidae), the San Joaquin Valley Tiger Beetle (*Cicindela tranquebarica* subsp., Carabidae), and the Mono Lake Hygrotus Diving Beetle (*Hygrotus artus,* Dystiscidae).

In a state as large, diverse, and unexplored as California, it is very likely that we have already lost species that we did not even know existed. Undoubtedly there are more beetles on the brink of extinction because of habitat loss, but many are unknown to scientists and land managers. Our lack of knowledge is compounded by limited resources for taxonomic and biodiversity studies, making it extremely difficult to properly assess and protect all the species that require the immediate attention of conservation biologists and governmental agencies.

Only a few California beetles have been afforded governmental protection. Many other species (see the section "California's Sensitive Beetles") living in sensitive habitats, especially those inhabiting coastal and desert sand dunes, require further study to determine if they are in need of state or federal protection.

California's Endangered Beetles

The federal Endangered Species Act (ESA) of 1973, and its amendments, is administered by the United States Fish and Wildlife Service. The ESA is designed to protect species and subspecies of organisms threatened or endangered by human activities. It includes provisions that encourage the acquisition of habitat, the development and implementation of recovery and habitat conservation plans, negotiating conservation agreements with local agencies, and the establishment and management of local preserves.

Of the 11 species of American beetles afforded protection by the ESA, four occur in California. These beetles are protected from any attempt to "harass, harm, pursue, shoot, wound, kill, trap, capture, collect, or attempt to engage in any such conduct." In addition, any action that would modify or destroy the habitat of a beetle listed as threatened or endangered is considered a violation of the ESA. California beetles may also receive protection under the state's own California Endangered Species Act, but as of yet none has been listed as threatened or endangered.

The Ohlone Tiger Beetle (*Cicindela ohlone*, Carabidae) (10 to 12.5 mm) was just listed as endangered in 2001. This species is restricted to five known populations located on remnant patches of open native grassland found on coastal terraces in the midcounty region of Santa Cruz County. The adults are bright green dorsally with tints of bronze on the pronotum. Unlike many other California tiger beetles, the Ohlone Tiger Beetle is active very early in the year, from late January to early April. Having been discovered and described only recently, in 1993, the historic range of this species is unknown, and the details of its biology are only now being discovered. The threats to this species include habitat fragmentation by urbanization and degradations by the invasion of nonnative vegetation, including grasses (Poaceae), filaree *(Erodium)*, and trees such as eucalyptus *(Eucalyptus)*.

The Delta Green Ground Beetle (*Elaphrus viridis*, Carabidae) (6 mm) is a striking, metallic green beetle that was listed as threatened in 1980. Once thought to be extinct, this species was rediscovered in the Nature Conservancy's Jepson Prairie Reserve, near Dixon in Solano County. Additional populations are known on privately held lands nearby. Its historical range probably included much of the Great Central Valley but was drastically reduced by agricultural development. Remaining populations are still threatened by agricultural activities and grazing. Additional surveys may yet reveal other populations elsewhere in the Great Central Valley. The Delta Green Ground Beetle is associated with vernal pool habitats. It emerges in January as the pools fill with water. It feeds on small, soft-bodied insects and breeds in February and March. As the pools begin to dry up in May, the adults burrow into the soil and remain there until the following January.

The Mount Hermon June Beetle (*Polyphylla barbata*, Scarabaeidae) (20 to 22 mm) is known only from the Zayante sand hills of Mount Hermon, Ben Lomond, and Scotts Valley areas of Santa Cruz County. It was listed as endangered in 1997. This beetle is similar to other beetles of the genus *Poly-*

phylla, except the elytra have scattered erect hairs and its stripes are broken up into scattered clumps. The larvae live underground where they feed on roots. The Mount Hermon June Beetle may complete its life cycle in one year. The adult males emerge in summer and take flight in search of females at the entrances of their burrows by homing in on their pheromones. Adult males are commonly attracted to lights. The Zayante sand hills habitat is threatened by sand mining, urban development, fire suppression, and agriculture. Several other endangered plants and animals occur in this habitat. A species of beetle considered by many to warrant consideration for federally protected status, the Santa Cruz Rain Beetle (*Pleocoma conjugens conjugens,* Pleocomidae), also occurs in the vicinity.

The Valley Elderberry Longhorn Beetle (*Desmocerus californicus dimorphus,* Cerambycidae) (13 to 21 mm), a subspecies of a more widely distributed beetle, the California Elderberry Longhorn Beetle *(D. c. californicus),* was also listed as threatened in 1980. The elytra of the males are mostly red with two pairs of dark, oblong spots; those of the female are dark bluish with reddish margins. Their larvae bore in the branches of blue elderberry *(Sambucus mexicana)* and red elderberry *(S. racemosa)* growing along streams and rivers in the Great Central Valley. Of the streamside forests believed to exist throughout the region 200 years ago, only an estimated 10 percent remains. Today small, scattered populations of beetles are known along the Sacramento, American, San Joaquin, Kings, Kaweah, and Tule River systems. Efforts are underway to protect this species by planting elderberry bushes and restoring streamside forests. Adults are seldom encountered, but their presence at a site is inferred by the distinctive, oval emergence holes in elderberry branches. As is true for other inhabitants of the Great Central Valley, the decline of the valley elderberry longhorn borer is due to agriculture. Continued threats include the clearing of streamside vegetation for firewood and the construction of waterways to irrigate fields. Cattle grazing and pesticides also pose additional threats.

Forest Pests

In the wild, the activities of wood-boring beetles are essential to the health of forest systems, accelerating the decay of wood and recycling its basic components for use by other plants and animals. In forests that are managed for timber production, however, the activities of these beetles can kill thousands of trees annually, resulting in significant monetary losses. Severe outbreaks of pest beetles and other insects also can increase the hazard of fire, damage watersheds and wildlife habitats, lead to increased erosion and sedimentation of streams, and reduce the aesthetic value of California's national parks and forests.

Several beetle families contain species that are sometimes considered forest pests in California, such as the metallic wood-boring beetles (Buprestidae) and the longhorn woodborers (Cerambycidae). For example, in addition to being a pest of agricultural crops such as hops, grapes, and fruit and nut trees, the California Prionus *(Prionus californicus)* attacks the roots of several species of California forest trees. The most destructive forest species, however, are the ambrosia and bark beetles (Curculionidae). They destroy billions of board feet of lumber annually. They also attack and kill trees around forest homes, park buildings, and other heavily used areas in the mountains.

Ambrosia Beetles

Ambrosia beetles are so named because of their relationship with the fungi they feed upon. The females carry spores of ambrosia fungi in a special chamber called the *mycangium.* They inoculate the walls of the tunnels bored in trees with the spores, which soon grow to provide a continuous food supply for both adult and larval beetles. They never feed on the wood itself. The tunnels are formed by the adults and are uniform in diameter.

Ambrosia beetles attack weakened, recently dead or freshly felled trees and bore directly into the sapwood. Because of the requirements of the fungus, ambrosia beetles are very exacting in choosing a host tree. If the wood is too wet, the beetles are overcome by the rapid growth of the fungus. If the wood is too dry, the fungus fails to thrive and the beetles soon starve. Only moist, unseasoned wood is prone to attack by ambrosia beetles. Their small round tunnels, called pinholes or shotholes, are often stained black, blue, or brown by fungal growth. Both the tunnels and stains seriously degrade the value of the lumber, resulting in significant financial losses.

A scolytine ambrosia beetle, the large California Oak Ambrosia Beetle *(Monarthrum scutellare)* (3.5 to 4.1 mm) is dark brown. The male takes about three to four weeks to completely excavate the main gallery and nuptial chamber in the coast live oak *(Quercus agrifolia)*. The female bores up to four secondary galleries leading away from the nuptial chamber, sometimes with the assistance of the male. Mating occurs when the galleries have been completed. The female then carves niches along the sides of the galleries into which she lays 50 or more eggs.

Bark Beetles

Bark beetles are so named because most of them live and mine between the bark and wood of trees and shrubs. The most destructive species attack the trunk of living trees, whereas others live in branches, twigs, cones, or roots. They are the most destructive forest insects in California. Once established, they attract waves of additional beetles by emitting highly attractive pheromones. Normal populations of bark beetles are usually small and scattered, but epidemic outbreaks may develop over large areas, killing countless trees, which become potential fuel for incredibly destructive forest fires. Outbreaks generally occur when trees are already stressed because of drought or fire injury, although healthy living trees are sometimes attacked.

As a mated pair of bark beetles bore through the pine bark, a reddish tube of pitch may begin to form at the entrance of the hole. As tunneling continues, sawdust is pushed out through the entrance. Pitch tubes and wood dust are the earliest signs of beetle attack and distinguish their activities from that of other wood-boring beetles. The foliage of fatally attacked trees soon turns yellow, then red, and quickly dies.

After boring through the bark, some bark beetles begin tunneling between the wood and the bark to construct an egg-laying gallery. In some species the chamber is formed by a single pair of beetles, whereas in others a single male may have a harem of two or three females, each carving out her own gallery. The hatching larvae often mine shallow tunnels away from the brood chamber in distinct radiating patterns. The mines are quite small at first but increase in size and fill with frass as the larva grows. The distinctive patterns created by the brood chambers and larval mines (pl. 13) led to the name "engraver beetle" and are used to identify some bark beetles. The larvae pupate at the end of the tunnel. Emerged adults may remain at the end of the larval tunnels awhile before chewing their way out through the bark.

Members of the genus *Dendroctonus,* a name derived from two Greek words meaning "tree killer," are stout, cylindrical, reddish brown to black beetles that typically seek out weak, dead, or dying trees to lay their eggs. Some species attack healthy trees. *Dendroctonus* beetles work in pairs to bore through bark and carve the brood chamber between the bark and wood. The egg galleries are always packed with wood dust, except for the portion where the beetles are working.

Plate 13. The Fir Engraver (*Scolytus ventralis,* Curculionidae) (4 mm), is a major pest of fir forests in California. After hatching, the larvae mine shallow tunnels away from the brood chamber in distinct radiating patterns.

Plate 14. The Red Turpentine Beetle (*Dendroctonus valens,* Curculionidae) (5.7 to 9.5 mm). Members of this genus cause significant damage to trees every year in the forests of California. Photo by C. L. Hogue.

The largest species is the Red Turpentine Beetle *(Dendroctonus valens)* (5.7 to 9.5 mm) (pl. 14). This reddish brown beetle attacks all species of pine found within its range. The stout Mountain Pine Beetle *(Dendroctonus ponderosae)* (4 to 7.5 mm) is dark brown to black. It is the most destructive species of bark beetle in western North America, attacking lodgepole *(Pinus contorta)*, ponderosa *(P. ponderosa)*, western white *(P. monticola)*, and sugar *(P. lambertiana)* pines. The Western Pine Beetle *(Dendroctonus brevicomis)* (3 to 5 mm) is dark brown and causes extensive mortality of ponderosa pine throughout the western United States and Canada. The Jeffrey Pine Beetle *(Dendroctonus jeffreyi)* (8 mm), is similar to the Mountain Pine Beetle but feeds only on Jeffrey pine. It can be a serious pest in old growth stands and timber-producing areas of northeastern California. It normally breeds in scattered mature trees, but it also attacks lightning-struck trees and trees recently toppled by high winds. During outbreaks they can kill groups of 20 to 30 trees regardless of age or health. They are sometimes found with the California Flatheaded Borer (*Melanophila californica*, Buprestidae) or the Pine Engraver *(Ips pini)*. The Douglas-fir Beetle *(Dendroctonus pseudotsugae)* (4.4 to 7 mm) is a rather hairy, dark brown to black species with reddish elytra.

Engraver beetles of the genus *Ips* are bark beetles that attack mostly pine and spruces. The name *Ips* is derived from the Greek word meaning "worm," inspired by the appearance of their tunneling activities. Mature larvae pupate at the blind end of the larval mine in a specially constructed cell. The shiny adults are reddish brown to black and cylindrical in shape. They range in length up to 6.5 mm. A distinguishing feature of this genus is the concavity at the rear of the wing covers, the margin of which is trimmed with three to six irregularly shaped spines. The number of spines, the host tree, and the location and pattern of the galleries are used as practical aids in identifying these beetles.

The first evidence of an *Ips* attack is the fine yellowish or reddish sawdust accumulating in bark crevices, in piles around entrance holes, or on the ground. Pitch tubes are seldom formed around the entrance holes. The foliage of infested trees turns from green to yellow, yellowish brown, or reddish brown. The tunneling activities of *Ips* beetles are easily observed by peeling the bark from an infested tree.

Outbreaks of *Ips* in standing, healthy trees are usually sporadic and of short duration. Control measures seldom reduce the damage caused by the beetles. To prevent situations favorable to damaging outbreaks, pile and burn cut wood before beetles can emerge from the wood. Another method is to scatter the wood in the sun to dry it out quickly, making it unsuitable for beetles to develop. Prompt salvage of wood downed by wind or storms also lessens the likelihood of *Ips* outbreaks. Occasionally, infested trees around homes are cut down and burned or treated with chemicals to stop outbreaks of engraver beetles.

The reddish brown Pine Engraver *(Ips pini)* (3.5 to 4.2 mm) is found across North America. One of the most common of the bark beetles, this species can become a serious pest in almost any species of pine in California, especially ponderosa, Jeffrey, and lodgepole pines. They normally attack downed wood or limbs already killed by *Dendroctonus* but frequently develop large populations that attack healthy, living trees. The California Fivespined Ips *(Ips paraconfusus)* (4 to 4.5 mm), attacks all species of pines in its range, especially ponderosa and sugar pines. This reddish brown to black species prefers smaller trees but will infest the tops of larger individuals. The Emarginate Ips *(Ips emarginatus)* (5.5 to 7 mm) is commonly associated with the Mountain Bark Beetle and Jeffrey Pine Beetle, attacking ponderosa, lodgepole, and western white pines. The shiny, dark brown adults engrave long, parallel egg galleries that connect at various points. The two- to four-foot long galleries run up and down the length of the tree.

Urban Assault Beetles

Long before humans arrived on the scene, beetles were already nibbling on scattered seeds and grains or scavenging rotting fruits and decaying flesh. The moment we began to store and process these very same materials for our own use, our relationship with these beetles was forever cemented. They quickly adapted to living in our food stores and soon became unwelcome pests. Many of these beetles are now cosmopolitan, residing in factories, mills, warehouses, grocery stores, and pantries around the world, dispersed by commerce and other human activities.

Pantry beetles are particularly annoying to homeowners because of their small size and ability to infest dried foods and spices that are not properly stored. They fly into buildings from outside sources such as nearby rodent, bird, or insect nests. They also hitchhike into homes on old furniture, rugs, drapes, bedding, and other materials of plant or animal origin. Once inside they take up residence in dark secluded corners, living on organic debris accumulating in cracks and crevices. Carefully tucked away in their hiding places, pantry beetles are poised to attack uninfested items in our homes as soon as they are brought in from the grocery store.

California Pantry Pests

The death watch beetles (Anobiidae) include several important pests of stored products. The bane of cigar aficionados, the Cigarette Beetle *(Lasioderma serricorne)* (2 to 4 mm), is also a serious pest of spices, legumes, grains, and cereal products. This oval beetle is light reddish brown and clothed in short, dense hairs. The antennae are saw toothed in appearance. The Drugstore Beetle *(Stegobium paniceum)* (2 to 3 mm) is cosmopolitan and attacks spices, herbs, legumes, biscuits, and candy. The beetle is straight sided and light reddish brown and is clothed with short, bent, and scattered erect setae. The last three antennal segments are gradually expanded to form a loose club. The Brown Spider Beetle *(Ptinus clavipes)* (2.3 to 3.2 mm) is widely distributed in California. It is a scavenger throughout the home, feeding on books, feathers, rodent droppings, sugar, legumes, seeds, animal feeds, and all kinds of grains. Spider beetles are so named because of their small, round bodies and six spidery legs. A pair of long antennae completes their eight-legged appearance. They range in color from pale to blackish brown.

Members of the genus *Necrobia* (Cleridae) are primarily scavengers of dry carrion and other dried organic matter. The Redlegged Ham Beetle *(Necrobia rufipes)* (3.5 to 7 mm) (pl. 15) is a cosmopolitan beetle that infests dried and smoked meats. It is a unicolorous dark metallic blue or green with reddish brown legs. Most of the damage is done by the larvae,

Plate 15. Adults and larvae of the cosmopolitan Redlegged Ham Beetle (*Necrobia rufipes,* Cleridae) (3.5 to 7 mm), infest dried and smoked meats, as well cheese, hides, dried carrion, and Egyptian mummies.

which prefer boring into the fattiest portions. They also attack cheese, hides, and dried carrion, such as old road kills and Egyptian mummies.

One of the most important pests of dried fruits is the Dried Fruit Beetle (*Carpophilus hemipterus*, Nitidulidae) (3 mm). It infests figs, dates, and occasionally raisins. This dull or shiny black beetle has shortened elytra, with each elytron marked with one large and one small yellowish brown spot. The full-grown larva is white or yellowish in color and may reach 7 mm in length. The complete life cycle may take as little as 15 days.

The omnivorous Sawtoothed Grain Beetle (*Oryzaephilus surinamensis*, Silvanidae) (3 mm) infests cereals, bread, pasta, nuts, cured meats, sugar, and many other products. Dried fruits, especially raisins, are usually infested only after they have been stored for long periods of time. The beetles' small, slender, flattened bodies permit their passage through narrow cracks and crevices, including the openings in poorly sealed food packages. The six broad, sawlike spines on each side of the prothorax easily distinguish these small brown beetles.

The Confused Flour Beetle (*Tribolium confusum*, Tenebrionidae) (3 to 4 mm) is considered one of the most important pests of food stored in supermarkets and homes. These small, reddish brown beetles infest legumes, shelled nuts, dried fruits, spices, chocolate, and even drugs and museum specimens. Unable to attack whole grains, the Confused Flour Beetle usually becomes a serious pest in flour mills. Although winged, it cannot fly. The sides of the prothorax are parallel, and the last four antennal segments form a loose club. The female may live two to three years, laying two or three sticky eggs per day in cracks or folds of flour sacks.

The Rice Weevil (*Sitophilus oryzae*, Curculionidae) (2 to 3.5 mm) is a nearly cosmopolitan pest of stored cereals, especially rice. It has a snout of moderate length and a dull, reddish brown body that is coarse in texture. The elytra often have four reddish yellow spots. The Grain Weevil (*S. gra-*

narius) is similar in size and appearance but usually infests wheat and barley products.

Clean Up and Clean Out

Several species of pantry pest beetles breed in food that has been spilled on shelves or become lodged in cracks, crevices, and corners. Periodic cleaning of these and other areas reduces their numbers. The only way to completely rid your home of pantry pests is to remove and destroy infested products. Carefully inspect any and all packages of cereal, dried food, nuts, flour, meal, pasta, chocolate, cocoa, soup mixes, spices, dry pet foods, and bird seed. Place suspect items in a clear, sealed plastic bag for later inspection to determine if there is beetle activity. When in doubt, throw it out!

Avoid storing foods for long periods by purchasing quantities of food that can be used quickly. Unused and uninfested materials must be stored in sealed tins or heavy plastic containers with tight fitting lids to thwart infestations by beetles and other pests.

Other Household Pests

Many species of skin beetles (Dermestidae) feed on a wide variety of materials of plant and animal origin, including treasured items in our homes. The cosmopolitan Furniture Carpet Beetle *(Anthrenus flavipes)* (2 to 3.5 mm) attacks upholstered furniture, hair padding, feathers, and woolens, as well as book bindings and brushes made of natural fibers. The golden brown larvae of the Museum Beetle *(A. museorum)* (2.2 to 3.6 mm) is nearly cosmopolitan. It feeds on furs, woolen materials, animal mounts, and dried insects. The adults are mottled to varying degrees with tan and gray scales. The Varied Carpet Beetle *(A. verbasci)* (2.5 to 3 mm) (pl. 16) is distributed nearly worldwide and is often found in granaries and flour mills. It is readily distinguished from all

Plate 16. The Varied Carpet Beetle (*Anthrenus verbasci,* Dermestidae) (2 to 3 mm) and other members of this genus, though frequently encountered on flowers, are very destructive pests of stored grains, wool fabrics, and insect collections.

other *Anthrenus* in North America by its long, narrow scales. The scales are usually white or yellowish and arranged in a zigzag pattern on a dark brown to black background. The larvae develop in a wide variety of foods and are pests in insect collections. They feed on the insect remains in spider webs and nests and apparently prey on the eggs of the Gypsy Moth *(Lymnatria dispar).*

Invaders from Down Under

Eucalyptus trees were originally imported into California from Australia about 140 years ago. Fast growing and adapted to life in a dry climate, eucalyptus trees quickly gained popularity as an evergreen ornamental. They were planted along the edges of agricultural fields as windbreaks and cultivated as a potential timber source; however, commercial efforts to use the lumber for railroad ties and milled wood proved unsuccessful.

For more than 100 years, California's eucalyptus trees were seldom attacked by insect pests or disease. Since the first trees were originally imported as seeds, all of their naturally occurring pests and pathogens were left behind. But their pest-free status began to change in the early 1980s with the discovery of several Australian imports.

Eucalyptus Longhorn Beetles

Eucalyptus longhorn beetles of the genus *Phoracantha* are native to Australia, where their larvae infest broken branches, cut logs, and trees stressed by injury or drought. They seldom attack healthy trees and are not generally considered to be serious pests in their native land. The beetles are very destructive in regions of the world where eucalyptus trees have been introduced, however, including parts of Europe, the Middle East, South America, and southern Africa.

The first *Phoracantha semipunctata* (12 to 30 mm) ever found in North America was collected in 1984 on dying eucalyptus trees in Orange County. The species has spread throughout California, killing thousands of trees. In June 1995, an even more destructive eucalyptus longhorn borer, *P. recurva* (pl. 17), was discovered at the University of California at Riverside. This new species spread quickly throughout southern California, where recent surveys have shown that it is replacing *P. semipunctata* as the most common species of eucalyptus longhorn beetle.

Phoracantha recurva strongly resembles *P. semipunctata* in size, shape, and color. The elytra of both beetles are dark brown, with cream to yellowish patches in clearly defined patterns. The elytra of *P. semipunctata* are mostly dark brown with a central cream-colored band divided by a dark zigzag line. In *P. recurva* the elytra are largely cream colored. The dark brown areas are mostly limited to a narrow strip along

Plate 17. The eucalyptus borer *Phoracantha recurva* (Cerambycidae) (12 to 30 mm) has spread throughout California, killing thousands of trees. Adults are attracted to fallen branches and injured or water-stressed trees.

the base and lower third of the elytra. The antennae of both beetles are longer than the body. The antennae of *P. recurva* are densely clothed with golden hairs, whereas those of *P. semipunctata* are bare.

The habits of both species are very similar. Adults are attracted to fallen branches and injured or water-stressed trees. After mating, females lay several batches of up to 30 yellowish eggs in cracks or crevices on the surfaces of branches and logs, or beneath loose bark. Freshly cut or fallen eucalyptus logs, as well as the branches of living trees, are also used. After several days, the larvae hatch and bore into the wood layer just beneath the bark, creating frass-filled, meandering tunnels as they feed. The width of the tunnels gradually increases as the larvae grow. The tunnels may extend about a meter along branches and trunks and are easily exposed by peeling away loose, dry bark. The foliage of trees infested with eucalyptus longhorn larvae wilts and dries but typically remains attached to the branches. Saplings may die within the first year after infestation, whereas mature trees may die within two years.

The whitish grubs reach a length of 30 mm or more when mature. In about six to eight weeks the larvae bore several centimeters into the sapwood to construct a pupal chamber. The pupal stage lasts up to 20 days. The entire life cycle, from egg to adult, takes about three to four months during summer but may last up to nine months for individuals that overwinter as larvae. Two generations are produced per season in southern California.

Adults emerge from circular exit holes 13 to 19 mm in diameter. The holes are sometimes marked with lines of dripping sap. They feed on eucalyptus flowers and are often found hiding beneath loose sheets of eucalyptus bark. Eucalyptus longhorn beetles are strong fliers and may travel miles from their emergence site to find suitable eucalyptus trees for mating and egg laying. Although the beetles prefer to attack blue gum *(Eucalyptus globulus)* stressed by lack of water or injury, recent studies indicate that even healthy trees are susceptible to attack.

Chemical sprays to control the larvae are ineffective because they cannot penetrate the wood. The best way to protect eucalyptus trees is to maintain their health. Regular deep irrigation minimizes water stress during the dry summer months. Prune and cut trees during winter and early spring when the adults are not active. Cover freshly cut wood immediately with tightly wrapped tarps or plastic sheets, and leave it for at least six months. Wood sealed in this manner prevents emerging adults from escaping and prevents its attractive odor from luring egg-laying females. Also, the heat built up in the wrapped wood can kill many of the larvae present inside the limbs and trunks.

University scientists are currently working to establish natural enemies of the eucalyptus longhorn beetles. The Australian parasitoid wasp (*Avetianella longoi,* Encyrtidae) lays its eggs on the eggs of the beetle. Hundreds of thousands of these tiny wasps have been released throughout California and are reported to have successfully reduced beetle populations in several parts of the state.

Other Eucalyptus Pests

Two more Australian beetles recently discovered in California have become important pests of eucalyptus. The first New World record of the Eucalyptus Tortoise Beetle (*Trachymela sloanei*, Chrysomelidae) (6 to 7 mm) (pl. 18) was in western Riverside County in 1998. Resembling a lady beetle (Coccinellidae) in shape, these brown and mottled beetles are now found throughout southern California, where they feed on

Plate 18. The Eucalyptus Tortoise Beetle (*Trachymela sloanei*, Chrysomelidae) (6 to 7 mm) is a recently established pest of eucalyptus trees in southern California.

red gum *(Eucalyptus camaldulensis).* The tarsomeres are broad and thick and are covered below with thick brushy pads. The caterpillar-like larvae are grayish brown with thick bumps of different sizes scattered on the body. Eucalyptus tortoise beetles lay their eggs in small batches in cracks or crevices on or beneath bark. The entire life cycle takes two or three months. Both adults and larvae are voracious leaf feeders on both blue and red gum, but it is not clear whether or not they pose a serious threat to the tree.

The Eucalyptus Snout Beetle (*Gonipterus scutellatus,* Curculionidae) (6 to 10 mm) was discovered in Summerland, Santa Barbara County, in May 1996. Heavy infestations of adults and larvae cause extensive damage to blue gum by defoliating entire sections of trees planted as windbreaks. Trees not killed are severely stressed and become susceptible to attack by eucalyptus longhorn beetles and disease. Adult weevils feed along the leaf margin, leaving in their wake a distinctive series of irregular notches. The larvae mine the tissues between the upper and lower surfaces of the leaf. Adults are dark to orange brown and have a short, broad snout. When viewed from above they have rounded projections on either side located just below the base of the wing cover. The larvae are sluglike and slimy in appearance and produce chains of stringy frass that cling to their bodies. Young larvae are yellowish with small black dots, whereas mature individuals are yellow green with numerous small black dots and a pair of dark stripes running the length of the body. Pupation takes place in the soil. The length of development from egg to adult is about three to six months, with two to three generations produced per year. Their distribution appears limited to Ventura, Los Angeles, and Santa Barbara Counties. Another tiny Australian wasp, the egg parasite *Anaphes nitens* (Mymaridae), was introduced into these areas and has successfully reduced Eucalyptus Snout Beetle populations.

Stay on the Lookout

With thousands of people and tons of goods arriving daily from all parts of the world, California is continually bombarded with foreign beetle invasions. Thanks to federal and state quarantines, the risk of exposure to exotic pests has been significantly reduced. Inspection and trapping programs are in place to ensure that any pest that managed to slip through quarantine is quickly discovered and eradicated before it can become established. The following two species have already demonstrated their ability to colonize parts of North America and both pose a serious threat to California's agricultural and horticultural industries.

Japanese Beetle

During the summer of 1916 a few plump, shiny reddish bronze and green beetles flanked with tufts of white setae were discovered in a nursery near Riverton, New Jersey. The white patches quickly distinguished this species from all others in North America. These beetles probably arrived as grubs in shipments of root stock imported from Japan. At that time there was little known about the Japanese Beetle *(Popillia japonica,* Scarabaeidae) (8 to 12 mm) (pl. 19), other than it was common in parts of Japan, where it was seldom considered a pest. Similar scarab beetles, however, had caused great damage when they were accidentally introduced in other parts of the world, so local authorities began their attempts to monitor and control the beetle.

Adult Japanese Beetles spread rapidly, attacking the leaves, flowers, and fruits of nearly 300 species of plants, including residential plantings and agricultural crops. The subterranean white grub is C-shaped and may reach a length of 25 mm when mature. It feeds on the roots of plants and

Plate 19. If the Japanese Beetle (*Popillia japonica,* Scarabaeidae) (8 to 12 mm) became established in California it could cause hundreds of millions of dollars damage annually to gardens and agricultural crops.

can destroy large areas of pasture, lawn, parks, and golf courses.

By the 1950s, Japanese Beetles occupied nearly the entire eastern third of the United States, mostly east of the Mississippi River. That Japanese Beetles could survive the more arid regions of the Midwest was extremely unlikely. Nor was it likely they would cross the insurmountable barriers of the Rocky Mountains, Great Basin Desert, and the Sierra Nevada to reach the agricultural and urban regions of California. Instead, they hitched rides on planes. Surrounded by turf, many eastern airports were breeding grounds for millions of beetles. On sunny summer days, Japanese Beetles would take to the air by the thousands in search of food, mates, and egg-laying sites. Some of these beetles would inadvertently fly into cargo holds, whereas others clung to the hair and clothing of passengers as they boarded flights to the west.

The very first Japanese Beetle recorded in California was caught in a trap near Los Angeles International Airport in June 1951. Two more specimens were trapped near civilian or military airfields in 1954 and 1956. In 1960, nearly 350 living and dead beetles were found in the passenger and cargo compartments of flights originating in Baltimore, Philadelphia, New York, Boston, and Pittsburgh.

On June 7, 1961, a state entomologist collected a Japanese Beetle feeding on flowers on the grounds of the state capitol building. As it turned out, this beetle was just the tip of the iceberg, representing California's first Japanese Beetle infestation. Intensive inspections and trapping soon produced hundreds of live beetles from the downtown Sacramento area. Using extensive applications of foliage sprays and ground treatments, the Japanese Beetle was declared eradicated in 1965, three years after the last live beetle was found. Two subsequent infestations were detected and eradicated in Balboa Park, San Diego, and Orangevale, near Sacramento.

If Japanese Beetles became established in California it would affect 50 percent or more of the grapes, roses, peaches, prunes, plums, asparagus, and apricots grown in the United States each year. The potential losses of just 5 percent in these seven crops alone could run into the hundreds of millions of dollars annually. Garden plantings and some street trees would also be affected. Nearly all commercially grown flowers would be susceptible to damage as well. Uninfested regions would quarantine the state's agricultural and horticultural exports, making it difficult, costly, or impossible for California growers to sell their produce.

With all this at stake, state agricultural authorities launched annual Japanese Beetle detection programs as a precautionary measure. Every year, between May 1 and August 31, thousands of winged funnel traps resembling green or yellow lanterns are deployed throughout the state on a grid-based arrangement. The traps are fitted with chemical lures promising the beetles both food and sex. The duped insects fall down a funnel and into a container where they wait for the state or county inspector who checks the traps weekly. The traps are concentrated in heavily populated urban areas, especially near airports where hitchhiking beetles are likely to disembark. Both commercial and cargo planes coming from Japanese Beetle–infested states are regularly inspected during summer.

Asian Longhorn Beetle

A native of China, the Asian Longhorn Beetle (*Anoplophora glabripennis,* Cerambycidae) (25 to 35 mm) (pl. 20) attacks a broad range of hardwood trees. It has created havoc in parts of China and Korea by infesting trees planted as windbreaks. It has also become a pest in plantations of poplar and willow used to make shipping crates. Untreated and infested lumber used to build shipping crates and wooden spools is responsible for the introduction of this conspicuous and destructive beetle into North America.

Plate 20. The Asian Longhorn Beetle (*Anoplophora glabripennis,* Cerambycidae) (25 to 30 mm) attacks a broad range of hardwood trees. Both adults and larvae have been intercepted in forest products shipped to other parts of the country, including California. Photo courtesy of James E. Appleby, University of Illinois.

It was first discovered in Brooklyn and Long Island, New York, during the fall of 1996. Some experts believe that the invasion of the Asian Longhorn Beetle began as much as 10 years earlier, when the insects hitchhiked in untreated wooden crates from China used to ship tools for a water treatment facility in Brooklyn. Imports of untreated wood from China were immediately banned, but the reuse of beetle-infected crates already in the United States may have spread the Asian Longhorn Beetles elsewhere.

The beetles later spread to other New York communities where they threatened several species of street trees, including maples, elms, horsechestnut, mulberry, and black locust trees. In 1998, the Asian Longhorn Beetle was discovered in Chicago wandering about a load of firewood in the back of a pickup truck. This second infestation occurred independently of the New York introduction and remained undetected for several years.

Asian Longhorn Beetles also threaten the maple syrup and lumber industries, potentially resulting in millions of dollars worth of damage. Although both adults and larvae have been intercepted in forest products shipped to other parts of the country, including California, easy identification and cargo fumigation has so far kept the beetle from becoming established elsewhere.

Asian Longhorn Beetles are large, shiny black beetles with about 20 white spots on their elytra. The black antennae are ringed with white and extend well beyond the length of the body. Their broad, flattened feet are black with a whitish blue upper surface. Adults emerge in spring and summer from branches, trunks, and exposed roots to mate and lay eggs, leaving round exit holes about 10 mm in diameter. Large piles of coarse sawdust accumulating around the base of the trees or where branches meet the main stem is evidence of beetle emergence.

Female Asian Longhorn Beetles chew out places on or under the bark to lay 30 to 80 eggs, leaving darkened oval to round wounds in the bark. The larvae of most native longhorn beetles (Cerambycidae) rarely harm living wood, preferring instead to eat dead wood and thus function as important decomposers in forest systems. On the other hand, Asian Longhorn larvae burrow into healthy living wood to feed. Individual trees are often attacked repeatedly by successive generations, whose combined tunneling efforts result in the death of the tree.

Trees infested with Asian Longhorn Beetles are cut down, chipped, and burned. In three years alone, more than 3,800 trees in New York City were destroyed in this manner, whereas Chicago lost nearly 1,000 trees. American scientists are working with their Chinese counterparts to find an effective and less destructive method for controlling Asian Longhorn Beetles. For example, traps with synthesized pheromones and other attractive chemicals are being investigated to lure beetles from their protective hiding places as part of an early detection system.

Sanitation Engineers

Dung beetles have long captured the imaginations of entomologists and casual observers alike, as they laboriously prepare animal wastes as food for themselves and their young. But ever since the disappearance of most of the resident large mammals 10,000 years ago, California has experienced a general lack of large piles of dung. As a result, California has only a few native dung scarab beetles (e.g., *Canthon simplex* and *Liatongus californicus*) that are capable of handling the big, juicy droppings produced by the state's cattle. Furthermore, neither of these beetles is particularly prevalent in cattle-producing areas. Several species of European *Aphodius* dung beetles (Scarabaeidae), long established in California, are commonly found in horse and cattle dung, but they prefer to breed in the dung pad rather than to bury it.

As cattle dung began to pile up, so did the numbers of cattle pests, especially flies. The Face Fly *(Musca autumnalis)* and the blood-sucking Horn Fly *(Haematobia irritans)* both breed in dung. Blood-sucking Horn Flies can annoy cattle to such an extent that they can cause losses of millions of dollars annually through lost milk production and weight loss. Face Flies spread pinkeye, costing farmers millions of dollars in treatment and losses of stock caused by blindness.

Dried-out dung pads pose another major problem. They can last up to three years before being trampled to bits or eaten by termites. These so-called prairie pies not only tie up valuable nutrients, they smother edible new growth yet encourage the growth of rank herbage that cattle refuse to eat. As they accumulated daily across the state, California risked losing millions of acres of pasturage annually.

The Australians were dealing with a similar problem and began importing dung scarabs from Africa to remove the cow pads. It was well known to researchers that dung beetles compete directly with pestiferous flies by eating the dung and burying it as food for their larvae. The rapid burial of dung reduces the number of egg-laying sites for flies and destroys the eggs of potentially infective intestinal parasites developing in the dung as well. Dung burial also reduces the amount of pasturage lost due to the growth of rank herbage surrounding old pads left on the ground surface. Dung burial aerates and fertilizes the soil, possibly increasing the amount of available pastureland by as much as 10 percent.

The California Dung Beetle Project

State agricultural authorities began to purposefully introduce additional species of exotic dung beetles in 1973 when the first few beetles were reared in a corner of a greenhouse at the University of California at Davis. The initial breeding stock was received from the quarantine facilities of the Commonwealth Scientific and Industrial Organization of Australia and from the United States Department of Agriculture's dung beetle project in Texas.

The California Dung Beetle Project quickly expanded to the mass rearing of four species of exotic dung beetles. During the summer of 1974 two species were released at the University of California's Sierra Foothill Range Field Station. Additional groups of beetles were placed in large screened cages in Alameda, Humboldt, Monterey, and Ventura Counties to determine their ability to survive and reproduce under a wide variety of climates. In 1975, 7,000 beetles representing two species were released into pastures and rangelands. The number of dung beetles released grew to more than 38,000 in 1976 and 250,000 in 1977. After releasing a total of nearly 680,000 dung beetles on more than 20 ranches throughout the state, federal funding for the program was curtailed. Undaunted, enthusiastic farmers were eager to enlist the dung removal services of the beetles and implemented their own release program by transporting buckets of beetle-infested manure wherever they were needed. Four species of intentionally introduced exotic dung beetles are currently established in California.

Onthophagus gazella (8 to 13 mm) is an African-Asian species that was first introduced into North America with releases in Texas in 1972. Since then it has become common throughout the southern United States. It was released at various sites in central and southern California and is now well established in the coastal plains of Ventura County southward, the Coachella Valley, and the coastal chaparral regions of Orange and San Diego Counties. These light to dark brown beetles fly to fresh pads of horse and cow dung from dusk to dawn and are readily attracted to lights. Males have a pair of short straight horns on their head, whereas the females are hornless. *Onthophagus taurus* (8 to 10 mm) (pl. 21) is a small species native to southern Europe and Asia Minor. It is nearly all black, sometimes with a dark greenish sheen. It is widely established in California, especially in urban areas where it is often attracted to dog feces. Major males have a long pair of

Plate 21. *Onthophagus taurus* (Scarabaeidae) (8 to 10 mm), native to southern Europe and Asia Minor, is widely established in California. It is sometimes common in urban areas where it feeds on dog feces. Major males have a long pair of curved horns sweeping back over the pronotum.

curved horns sweeping back over the pronotum. Both exotic *Onthophagus* species bury their dung balls in chambers dug next to the dung pad.

Onitis alexis (12 to 20 mm) is an oblong, robust species native to Africa. It also occurs in southern Europe and has been successfully introduced into Australia. It is easily distinguished from the other exotic dung beetles in California by its large size, dark brown or coppery green pronotum with two distinct pits centrally located on the rear margin, and distinctly grooved light brown elytra. Both the male and female lack horns on their head. The males have a single hooklike spine located on the inside of each hind femur, a feature absent in females. They are active at dusk and prefer dung that is two days old or less. *Onitis* is established along the coastal plain of southern California.

Euoniticellus intermedius (7 to 9 mm) is light brown in color with distinctive symmetrical markings on the pronotum. The male has a single, blunt, curved horn on the head. A native of sub-Saharan Africa, *E. intermedius* searches for fresh dung pads during the day. Adults live one or two months, during which time the female can lay more than 100 eggs. The nest structure varies depending on the availability of moisture. During the wetter parts of the year the nest of *E. intermedius* may consist of a main burrow with several branched galleries, each provided with a pear-shaped ball of dung. Under drier conditions the brood balls are placed immediately below the pad to take advantage of its available moisture.

California Beetles and the Birth of Modern Biological Control

In spite of such formidable obstacles as mountains, deserts, occasional unseasonable frosts, and periodic droughts, California rapidly became a powerhouse of American agriculture. The state's modern agricultural heritage is rooted in the early farms and ranches established by the settlers who remained after the Gold Rush in the mid-1800s. Previously, the native Americans in the region grew little food, whereas the Spanish maintained only small gardens at most of their missions established throughout much of the state. The establishment of the first transcontinental railroad in 1869 opened new markets for California's crops and led to increased production. The introduction of nursery stock and ornamental plants from around the world to meet the demands of the east resulted in the introduction and establishment of many serious insect pests.

The Spanish first brought citrus fruit to California. The first sizable grove was located at the San Gabriel mission, located just north of Los Angeles. Commercial production of oranges was well underway by the time of the Gold Rush in the 1850s. In 1868 a particularly pesky relative of aphids (Aphididae) and mealy bugs (Pseudococcidae) known as the Cottony Cushion Scale (*Icerya purchasi,* Margarodidae), was first noticed on acacias imported from Australia in Menlo Park just south of San Francisco. By 1885, it had become a serious pest of citrus orchards throughout the state, threatening the entire fledgling industry. C.V. Riley, chief entomologist of the United States Department of Agriculture, was aware of the scale's Australian origins. In 1887 he sent A. Koebele, a naturalized German immigrant, to Australia to look for and collect natural enemies of the scale. Koebele sent three shipments in late 1888 and early 1889 to D. Coquillet in Los Angeles. These shipments included several species of ladybugs (Coccinellidae), including 129 specimens of the Vedalia Beetle *(Rodolia cardinalis)* and were the first exotic insects introduced into North America for the purpose of controlling an insect pest. Coquillet placed all the Vedalia on one small, tented citrus tree infested with scale near Los Angeles.

Within a few months the scales were gone, and the Vedalia Beetles were soon released among the surrounding trees. An additional 385 beetles were imported from Australia and released directly into the orchards in early 1889. In just two months, over 10,000 beetles were collected and distributed among citrus farmers throughout southern California. Within one year the citrus crop increased nearly threefold, marking the complete recovery of the citrus industry. The Vedalia Beetle was hailed as a miracle of science. Along with another Australian import, the parasitic fly (*Cryptochetum iceryae,* Cryptochetidae), the Vedalia Beetle has kept the Cottony Cushion Scale in check now for more than 100 years. Descendants of these and additional Australian beetles raised in California have been shipped throughout the world to successfully control outbreaks of Cottony Cushion Scale.

The immediate success of the introduction of *Rodolia* led to the now well-established method of introducing predatory insects, parasites, and parasitoids to control both animal and plant pests. Between 1891 and 1892 more than 45 other species of ladybugs were introduced into the United States, but only a few became established. The basic requirements and biologies of these introduced species were little known or ignored. With enthusiasm for ladybugs at an all-time high, native California ladybugs were shipped abroad to control pest species. Thousands of Convergent Ladybugs *(Hippodamia convergens)* were shipped to South Africa but failed to adapt to the climate of their new home.

Tinkering with Nature

Pest control of any kind has its risks and may adversely affect other species or their habitats. By design, classical biological control is based on the fact that the introduction of alien organisms disrupts established populations. Extensive tests are usually conducted to determine that these introduced organisms do not adversely affect nontarget organisms. In some cases though, biological control introductions have become part of the much larger problem of alien species invasions, which are recognized as a major factor in local and species-wide extinction.

The impact of exotic dung scarabs on native California dung beetle populations is currently unknown because a comprehensive survey has never been undertaken. Officials with the now-defunct California Dung Beetle Project had declared that native dung beetles were either not present in the proposed release areas or that their role in cattle dung burial was insignificant. Prerelease studies focused on the abilities of exotic beetles to survive in a variety of California climates. Once released into the environment, the only concern was that the beetles would become a permanent, self-perpetuating type of pest management capable of burying more dung as they expanded their range.

No studies were performed to determine the impact of exotic dung beetle populations on native species living outside cattle-producing areas. As the exotic beetles disperse into the foothills of the western Sierra Nevada and elsewhere, they will undoubtedly reduce the amount of dung available to native species such as *Canthon* and *Liatongus.* After 25 years with exotic dung beetles in California, it would seem desirable to review the benefit of exotic dung beetles and their impact on native species. This information would be useful in weighing the pros and cons of importing more exotic beetles as biological control agents in the future.

CHAPTER 6

COMMON AND CONSPICUOUS FAMILIES OF CALIFORNIA BEETLES

Of the 131 families of beetles known in North America, all but 13 are found in California (see "Checklist of North American Beetle Families"). The characteristics and habits of 23 of the more common or conspicuous families in the state are reviewed below.

Ground Beetles and Tiger Beetles (Carabidae)

Ground beetles and tiger beetles are among the most common and frequently encountered beetles. They are quite diverse and make up California's third largest family of beetles, behind the rove beetles (Staphylinidae) and the weevils and bark beetles (Curculionidae). With long legs and powerful jaws, ground beetles and tiger beetles are perfectly adapted for hunting and killing insect prey. Both larvae and adults are considered largely beneficial because they prey on caterpillars, grubs, grasshoppers, snails, and other small animals that are considered pests. A few species may occasionally consume fruits and seeds. Most tiger beetles hunt during the warmest part of the day, whereas many ground beetles and some tiger beetles hunt at dusk or at night. Ground beetles and tiger beetles run swiftly when disturbed and often release a stream of noxious fluid from their anus, sometimes with amazing accuracy. Pydigial glands located at the tip of their abdomens produce a variety of hydrocarbons, aldehydes, phenols, quinones, esters, and acids that combine to produce foul-smelling and sometimes caustic fluids.

The females of several genera of ground beetles *(Chlaenius, Brachinus, Pterostichus,* and *Calosoma)* lay their eggs in specially constructed cells made of mud, twigs, and leaves. The larva molts three times before becoming a pupa. Ground beetle larvae are generally active predators that crawl on the ground or lurk beneath bark. Tiger beetle larvae live in vertical burrows in fine or sandy soils. The rather long, cylindrical body of a tiger beetle larva is anchored inside the burrow with abdominal hooks. Sensing the vibrations of an insect near the burrow's entrance, the larva lunges outside the burrow toward its target, often exposing up to half its body length. If successful, it grabs its prey with large, scythelike jaws and drags it back into the burrow, where the larva can dine in relative safety.

Tiger beetles, with their distinct shape and bulging eyes, were once placed in their own family, Cicindelidae, but are now considered a subfamily of Carabidae. Their shiny bright colors and distinct markings help them blend into their background, making it difficult for potential predators to track them. The body patterns of some species may mimic the appearance of blistering or stinging insects or help them to regulate their body temperature. Robber flies (Asilidae) readily capture and eat adult tiger beetles, whereas several birds prey on both larvae and adults.

CALIFORNIA FAUNA: 677 species and subspecies in 91 genera.

IDENTIFICATION: Ground beetles (pl. 22) are small to large elongate beetles with flattened, hairless bodies tapered at both ends. Ground beetles are usually uniformly dark and shiny and sometimes brownish or pale. Some species are metallic or bicolored or tricolored or with the elytra brightly marked with patterns of yellow or orange. Tiger beetles (pl. 23) often have bold white or yellow markings on otherwise metallic elytra. They range in length up to 34 mm. The head is directed forward, with 11-segmented threadlike or beadlike antennae. The pronotum is usually narrower than the elytra. In ground beetles the sides of the pronotum are sharply margined, less so in tiger beetles. The scutellum is visible except in the round sand beetles, of the genus *Omophron*. The elytra are always present, completely cover the abdomen, and are fused in flightless species. The elytral surface is smooth, pitted, or grooved. Flight wings range from fully developed to absent. The flattened legs of ground beetles and tiger beetles adapt them for various activities, including walking, running, and digging.

Plate 22. The Common Black Calosoma (*Calosoma semilaeve,* Carabidae) (20 to 27 mm) is extremely numerous in some years in the Great Central Valley and Mojave Desert. It is sometimes attracted to lights in large numbers.

Plate 23. Tiger beetles, such as *Cicindela californica* (Carabidae) (12 mm), are typically brightly colored, often metallic, and have bulging eyes that they use to actively seek out and hunt their prey.

The tarsal formula is 5-5-5, with claws equal and simple, saw toothed, or comblike. The abdomen has six segments visible below, except for the genus *Brachinus,* which has seven or eight. The first abdominal segment is partially covered by the bases, or coxae, of the hind legs.

The yellow, brown, reddish brown, or black larvae are almost cylindrical or slightly flattened. The antennae usually have four segments. Most species have six simple eyes on each side of the head, but a few do not. The legs are usually six segmented. The 10-segmented abdomen has projections on segment nine.

SIMILAR CALIFORNIA FAMILIES:

- false tiger beetles (Salpingidae, Othniinae)—small (5 to 9 mm); antennae clubbed
- darkling beetles (Tenebrionidae)—tarsi 5-5-4; antennae sometimes gradually clubbed
- comb-clawed beetles (Tenebrionidae, Alleculinae)—tarsi 5-5-4; claws saw toothed
- narrow-waisted bark beetles (Salpingidae)—tarsi 5-5-4

Whirligig Beetles (Gyrinidae)

Whirligigs are readily distinguished from other beetles by their flattened bodies, short, paddlelike legs, and extraordinary divided compound eyes. They are the only group of beetles that regularly use the surface film of water for support. Whirligigs live on the surface of ponds or along the edges of slow-moving streams. They are sometimes found in cattle tanks, canals, swimming pools, and even rain puddles. Moving about singly or in large groups, whirligigs scour the water's surface in search of food and mates. Recently emerged adults sometimes congregate by the dozens or hundreds in shady or sheltered spots in late summer and fall.

The rigid, streamlined bodies of whirligigs meet little resistance on the water. The divided compound eyes are especially adapted for life on the surface. The lenses of the upper and lower eye surfaces are specifically adapted for gathering images in the air and water, respectively. The raptorial front legs of both sexes are used to grasp dead or dying insects floating on the water. In males the raptorial front legs are also used to grapple with smooth and slippery mates. Propelled by their paddlelike middle and hind legs, whirligigs steer with their rudderlike abdomen. They are good fliers and are sometimes attracted to lights at night. Like other aquatic beetles, whirligigs breathe air stored beneath the elytra.

The name "whirligig" was bestowed upon these animals by virtue of their wild gyrations when disturbed. They occasionally dive beneath the surface for brief periods to escape from harm. Incredibly buoyant, whirligigs must cling to underwater objects to remain submerged. Adults also produce defensive secretions from the tip of the abdomen that have a pungent odor suggesting apples. This odor is responsible for the nicknames "apple bugs," "apple smellers," and "mellow bugs." The secretion is not only distasteful to fish, amphibians, birds, and other predators, but it may also serve as an alarm pheromone to alert nearby whirligigs of possible danger. This secretion also propels them through the water by lowering the surface tension. The whirligig is thought to ride on the wave of water molecules recoiling from the chemical.

Eggs are laid in rows, end to end, in clusters on submerged plants. The predatory larvae crawl about the bottom debris of ponds and streams in search of immature insects and other small invertebrates. They breathe directly through their body wall with the aid of threadlike projections jutting out from the sides of their bodies. Mature larvae pupate within a case constructed of sand and debris. Pupal cases are located on the shore or attached to plants above the water surface. Parasitic wasps are known to attack the pupae.

CALIFORNIA FAUNA: Nine species in three genera.

IDENTIFICATION: Whirligigs (pl. 24) are shiny or dull black beetles ranging in length from 3 to 15 mm. Their bodies are oval, flattened, and streamlined, with the combined margins of their head, thorax, and elytra forming a continuous outline. The head is directed forward and bears short, clubbed antennae consisting of eight to 11 segments. Their compound eyes are distinctly divided. The pronotum is equal to the width of the elytra. The scutellum may or may not be visible. The elytra are smooth and do not completely conceal the abdomen. The front legs are normally developed for grasping. The middle and hind legs are flattened and paddlelike. The tarsal formula is 5-5-5. The abdomen has six segments visible from below.

The white or yellowish brown larvae are elongate and slightly flattened. The head is distinctive and directed forward and bears a pair of four-segmented antennae. The abdomen

Plate 24. *Gyrinus plicifer* (Gyrinidae) (4.5 to 6.5 mm) is a member of the most commonly encountered genus of whirligig in California. *Gyrinis* is distinguished from other California genera by the presence of a visible scutellum. These beetles are frequently found in aggregations on the surface of quiet streams and pools.

has 10 segments, the last two of which are relatively narrow. A series of threadlike or feathery structures project from the sides of the abdomen and function as gills.

SIMILAR CALIFORNIA FAMILIES: The eyes, legs, and habits of whirligigs distinguish them from all other beetles.

Predaceous Diving Beetles (Dytiscidae)

Adult and larval diving beetles are predators and scavengers, consuming both invertebrate and vertebrate tissues. They occur in a wide variety of aquatic habitats, particularly along the edges of pools, ponds, and slow streams where emergent vegetation grows. A few species are specialists, preferring cold water streams, seeps, and springs, or other specialized bodies of water. Clear, gravel-bottomed streams, brackish waters, and thermal pools each have their own distinctive fauna. The diverse aquatic habitats of the wetter coastal and montane regions of northern and central California support the greatest diversity of diving beetles.

Predaceous diving beetles are descendents of terrestrial insects that evolved a number of adaptations for living in water. Their oval and flattened bodies are streamlined to reduce drag as they swim. The short, fringed hind legs move in unison, propelling the beetle through the water. The hind legs are placed well back on the body to increase speed and maneuverability. Diving beetles are quite awkward on land. Their legs are attached to plates tightly fused to the body and are incapable of moving up and down like their terrestrial counterparts.

Adults store air in a cavity beneath their wing covers, replacing it periodically by hanging their heads down from the water surface. This bubble also makes them quite buoyant. To control their buoyancy, diving beetles can reduce the size and carrying capacity of the cavity by expanding their abdomen into the space. Another method is to swallow water and store it as ballast in an expandable chamber at the end of their digestive system. The amount of water carried can be modified to regulate the beetle's position in the water.

While submerged, the covering of the exoskeleton is quite permeable to water. When the diving beetle has been out of the water for some time, however, it becomes somewhat waterproof. Beetles attempting to reenter the water often have difficulty in submerging and can become trapped on the surface film. Using the surface film as a foothold, struggling beetles usually manage to yank themselves below the water. To facilitate reentry into the water, the glands located at the tip of the abdomen produce chemical wetting agents called surfactants. When spread over the body surface, the surfactants make the exoskeleton more permeable to water and aid the beetle's efforts to become submerged.

The life histories of most California diving beetles are unknown. In general, the eggs are laid on moist soil or attached to or inserted into aquatic plants. The larvae are predaceous, and those of the larger species are sometimes called water tigers. Like the adults, many larvae obtain their oxygen from the surface where it may be stored internally in their tracheal trunks. Young larvae have closed spiracles along the sides of the body, but as they mature the spiracles open in preparation for breathing on land. Larvae of some species obtain oxygen from the water through their cuticle, and a few have gills. The larvae molt three times before pupation. Mature larvae leave the water to construct pupal chambers beneath streamside objects. Adult diving beetles are quite capable of flight and take to the air to colonize new bodies of water, occasionally taking up temporary quarters in swimming pools. They also fly to locate overwintering sites or to leave habitats that have dried up or become otherwise unsuitable.

The study of all water beetles is useful because they are long-lived and their presence or absence can serve as an indi-

Plate 25. The Yellow-spotted Diving Beetle, or Sunburst Diving Beetle, (*Thermonectus marmoratus,* Dytiscidae) (10 to 15 mm) is found in the Peninsular Ranges in slow-moving arroyos with rocky bottoms. The bright, contrasting color pattern renders the beetles inconspicuous as they rest and swim over sun-dappled, gravelly substrates and breaks up their image as a beetle.

cator of environmental quality. It is possible to calibrate the environmental tolerances of each species and use them to gauge a range of conditions. Predaceous diving beetles are of particular use in small pools and springs that lack other suitable indicator species.

CALIFORNIA FAUNA: 157 species in 27 genera.

IDENTIFICATION: Adult predaceous diving beetles (pl. 25) are oval and streamlined beetles ranging in length up to 33 mm. They are usually reddish brown to black or pale, with or without distinct markings. The head is directed forward and has beadlike antennae consisting of 11 segments. The pronotum is widest at the base. The scutellum is visible. The elytra are usually smooth and polished but may be pitted, sparsely hairy, or grooved, and completely conceal the abdomen. The tarsal formula is 5-5-5. The claws are equal or unequal in size and are not toothed. The abdomen has six segments visible from below.

The white, yellowish brown, or reddish brown to black larvae may be variously marked with yellow or black and are long, spindle shaped, or somewhat flattened. The head is directed forward and marked with a distinctive Y-shaped seam. The antennae are four segmented, and there are usually three or fewer simple eyes on either side of the head. The legs are six segmented. The abdominal segments are roughly equal in length. The last segment is tipped with one- or two-segmented projections.

SIMILAR CALIFORNIA FAMILIES:

- whirligig beetles (Gyrinidae)—eyes divided; antennae clubbed
- crawling water beetles (Haliplidae) have a small head and greatly expanded hind coxae
- burrowing water beetles (Noteridae)—scutellum not visible
- water scavenger beetles (Hydrophilidae)—antennae clubbed; mouthparts (maxillary palps) long; underside flat

Water Scavenger Beetles (Hydrophilidae)

Water scavenger beetles live in ponds, streams, and lakes with an abundance of plants or organic debris and are among the first arrivals in new aquatic habitats. They are particularly abundant along the vegetated edges of standing bodies of water and slow-moving streams or springs but are seldom found in brackish water. Like other aquatic beetles, water scavengers are attracted to the shiny surfaces of cars and even blue tarps. They are sometimes confused with predaceous diving beetles (Dytiscidae) but are easily distinguished by their humpbacked appearance, swimming style, and method for acquiring air bubbles. They also possess long, threadlike mouthparts that are sometimes mistaken for antennae.

The common name of this family is a bit of a misnomer. Water scavenger beetles live in a wide variety of habitats, including wet sand, rich organic soil, moist leaf litter, fresh mammal dung, or extremely decayed animal carcasses, as well as in water. The larvae are almost always predators, whereas the adults may be vegetarians, omnivores, or occasionally predators or scavengers. Predaceous species feed on a variety of animal foods, including snails and other small invertebrates, whereas omnivorous species add spores, algae, and decaying vegetation to their diets.

Aquatic water scavenger beetles capture and store a bubble of air along the ventral surface of their thorax and in the space under their elytra. They swim to the surface headfirst and break the surface with their antennae, which is in contrast to the predaceous diving beetles, which break through the surface with the tip of their abdomens. A makeshift funnel is formed by the dense velvety setae on the water scavenger's antennae, mouthparts, and thorax, connecting the surface air with the thorax and subelytral space. Air is exchanged by the pumping action of the elytra and abdomen.

Although many aquatic water scavenger beetles are good swimmers, some are slow and awkward. For better or worse, water scavengers propel themselves through the water by moving their legs in an alternate fashion, unlike predaceous diving beetles, which move their legs in unison. The setae on the legs of both adult and larval water scavenger beetles may or may not enhance their ability to swim. In at least one genus, *Berosus,* rows of hairs on the legs are repeatedly drawn through the air bubble in an apparent effort to increase aeration. This strategy might be especially useful for beetles living in stagnant bodies of water with low oxygen content. These hairs may also be part of a grooming system associated with special pores covering the abdomen.

Many aquatic species are capable of stridulating, producing sounds by rubbing their elytra and abdomens together. These sounds are emitted when the beetle is handled or under attack and is apparently a deterrent against predators. Sound production may be used by some species as a part of their courting behavior.

Up to 100 or more eggs are laid in a protective silken case that varies from a loose submerged pouch to a floatation device fitted with a sail-like breathing tube. The case is usually attached to the substrate or to vegetation and perhaps serves as a means of preventing the drowning of the eggs.

Larval water scavenger beetles are carnivorous, feeding upon other invertebrates. The growth rate is dependent upon temperature and the availability of food. They molt three times before pupation. Mature larvae leave the water to construct a pupal chamber of mud near the shore. The chamber is either buried in the soil or tucked beneath an object. The pupa is neatly suspended within the cell by strategically placed setae on its body. The newly emerged adult usually remains in the chamber until its body has darkened and hardened.

Fish, amphibians, reptiles, and aquatic birds all prey upon aquatic water scavenger beetles. At night, the beetles often take to the air in search of new habitats and are sometimes attracted to lights. During these nocturnal explorations they are subject to attack by still more birds and bats.

North American water scavenger beetles are of little economic importance, although larger species are reported to be pests in fish hatcheries. A few may be of some benefit as predators of mosquito larvae.

CALIFORNIA FAUNA: 71 species in 20 genera.

IDENTIFICATION: California water scavenger beetles (pl. 26) are broadly oval, distinctly arched on top, and flattened or sunken

Plate 26. The Giant Black Water Beetle (*Hydrophilus triangularis,* Hydrophilidae) (30 to 40 mm) is the largest beetle in this family and occurs east of the Sierra Nevada and along the Colorado River. Its shiny black elytra exhibit a greenish tinge. Adults are often attracted to lights at night, whereas the larvae may be important predators of mosquito larvae in rice fields.

underneath. They range in length up to 40 mm. The head is directed downward and bears gradually clubbed antennae with six to 10 segments. The last three segments form a variable-shaped club that is usually nested in the preceding cuplike segment. The maxillary palps often exceed the length of the antennae, but in terrestrial and semiaquatic species, they are usually equal in length or shorter. The pronotum is broader than the head and usually wider than it is long. The scutellum is visible. The elytra are widest at the middle and are broader at their base than the pronotum. The elytra are either smooth or rough and may be covered with rows of small pits and completely conceal the abdomen. The tarsal formula is 5-5-5, 5-4-4, or rarely 4-5-5. The claws are generally simple but sometimes modified in the male. The abdomen usually has five, or rarely six, segments visible underneath. Water scavengers are usually black, black with brownish markings, or rarely greenish or with cream markings.

The gray or yellowish brown larvae are quite variable in shape and are somewhat flattened, long, cylindrical, spindle shaped, or cone shaped. Their bodies are sometimes marked with patterns of dark setae. The distinct head is directed forward and bears six or fewer pairs of simple eyes. The antennae are three or four segmented. The legs have five segments. The eight- to 10-segmented abdomen rarely has gill-like appendages on the sides. Paired projections present on the tip of the abdomen may have from one to three segments.

SIMILAR CALIFORNIA FAMILIES:

- predaceous diving beetles (Dytiscidae)—antennae threadlike; mouthparts inconspicuous; body not flattened underneath
- minute moss beetles (Hydraenidae)—abdomen with six or seven visible segments underneath
- minute flower beetles (Phalacridae)—small (1 to 3 mm); maxillary palps short
- riffle beetles (Elmidae)—small; legs long with large claws
- dung beetles (Scarabaeidae)—antennal club with flat segments

Clown Beetles (Histeridae)

The clown beetle family is a fascinating yet largely unknown family of beetles. Both adults and larvae are primarily carnivorous, preying on insects. Many species are found on dung and carrion or decaying plants, where they feed on fly eggs and maggots. Some species stake out sapping wounds on trees and attack visiting insects. Others are specialists that live in ant nests and scavenge or prey on all stages of the ants. A few ant nest dwellers are so specialized in their form and behavior that they are actually fed by ants. Other species living with ants feed on other insects and surplus food collected by their hosts. Still other species live in bird or mammal nests, especially those of rodents, such as pack rats *(Neotoma)* and ground squirrels *(Spermophilus).* A few (e.g., *Hister*) are found in sand at the base of coastal or desert dune plants where they probably feed on the larvae of scarabs (Scarabaeidae), weevils (Curculionidae), and flies. When threatened, adults pull their head inside and tuck their legs tightly beneath their shiny, round, and compact bodies. Of the approximately 3,900 known species, 435 are known from the United States.

The larvae are unusual among the beetles, molting only twice, including the molt to the pupal stage. The thoracic legs are small and not useful for walking. Instead, they apparently move by waves of contractions of the abdomen. They feed on liquids and must digest their food outside of the body using digestive fluids.

As predators, some clown beetles are moderately beneficial. Species living in dung are probably of greater importance because they prey on pest flies infesting accumulations of animal waste in cattle pastures and poultry farms. Several dung-inhabiting species of *Hister* have been investigated as biological controls for the Horn Fly *(Haematobia irritans).* A few cylindrical species live in trees where they attack bark beetles (Curculionidae) inside their tunnels.

CALIFORNIA FAUNA: Approximately 140 species in 37 genera.

IDENTIFICATION: The majority of clown beetles (pl. 27) are shiny black, but a few species have distinct red markings or may be reddish or metallic blue or green. The head is directed forward or downward and typically bears prominent, sometimes very large mandibles. The 11-segmented antennae are elbowed and tipped with a compact three-segmented club that is often clothed in patches of sensory hairs. The elytra are distinctly grooved and short, usually appearing cut off, exposing the last one or two abdominal segments. The scutellum is visible. The tarsal formula is 5-5-5 or 5-5-4, with claws simple and sometimes unequal in size.

Plate 27. Nine species of *Saprinus* (Histeridae) are known from the United States. The most common species, *Saprinus lugens* (4.5 to 8 mm), is encountered in the foothills on carrion and dung. They are sometimes attracted to foul-smelling flowers that depend on carrion-visiting insects for pollination.

The larvae are generally long and cylindrical in shape and more or less round in cross section. The head, thoracic segments, and abdominal projections are thick and dark, whereas the rest of the body is thin and membranous. The head is directed forward and bears a single pair of simple eyes. The antennae are three segmented.

SIMILAR CALIFORNIA FAMILIES:

- false clown beetles (Sphaeritidae)—head concealed from above; elytra loosely covering abdomen
- shining fungus beetles (Staphylinidae, Scaphidiinae)—antennae clubbed; abdomen pointed
- spotted dung beetles (Hydrophilidae)—red and tan markings on elytra; antennae gradually clubbed; mandibles not distinct

Carrion Beetles (Silphidae)

Most carrion beetles feed primarily on decaying plant or animal material, whereas a few species prefer to eat living plants and can become minor garden pests. Although the larvae of carrion-feeding species appear to feed exclusively on dead animals, the adults of some of these species feed on carrion and insects, especially maggots. Carrion and burying beetles use smell to locate dead animals, using their antennae to detect hydrogen sulfide and some cyclic carbon compounds that are released as the carcass decays. Phoretic mites are often found wandering about the bodies of burying beetles found on carcasses or at lights. Their presence may benefit the beetles because they feed on fly eggs that might otherwise hatch and compete with the beetles for the carcass.

Adult pairs of burying beetles of the genus *Nicrophorus* demonstrate the most advanced behaviors of parental care known in beetles. They bury small animal carcasses to reduce competition with flies, ants, and other insects in an effort to provide food for both themselves and their larvae. Either the male or female initiates the carrion-burying process, during which time the arrival of a mate is likely. Mating occurs only after the carrion has been secured in a chamber beneath the ground. The carcass of a mouse, bird, or other animal is carefully buried and meticulously prepared by the adults. Feathers, hair, skin, and appendages are removed and the carcass is kneaded into a pear-shaped ball. The female then lays her eggs along the walls of a vertical chamber dug directly above the carcass. A conical depression created on the upper surface of the carcass receives droplets of regurgitated tissue deposited by both beetles. This accumulation of soup eventually serves as food for the newly hatched larvae. Adults stridulate by rubbing a pair of files located on their abdomen against the edge of their elytra, creating a scraping sound. Stridulation occurs

Plate 28. The Black Burying Beetle (*Nicrophorus nigrita,* Silphidae) (13 to 18 mm) lives along the Pacific Coast, from British Columbia southward to Baja California. The nocturnal adults are active from February through November in coastal forests and open habitats.

during stress, mating, confrontations with other beetles and is used to communicate with the larvae. The larvae receive parental care throughout their development, and only upon pupation of their young do the adults voluntarily leave the brood chamber.

CALIFORNIA FAUNA: Eight species in four genera.

IDENTIFICATION: California carrion beetles (pl. 28) are slightly to strongly flattened and black in color, sometimes with orange or reddish orange markings on the elytra. They range in size from 7 to 22 mm in length. The head is directed forward. The antennae are nine or 10 segmented with gradually or abruptly clubbed antennae that are velvety in appearance. The pronotum is broader than the head with strong margins. The scutellum is visible. The elytra are sometimes truncate, exposing one or more segments of the abdomen. The surface of the elytra is rough or smooth, sometimes with three raised ribs, but never grooved. The tarsal formula is 5-5-5, with claws equal in size and simple. The abdomen has six, or rarely seven, segments visible below.

The larvae of California carrion beetles are elongate and flattened. The segments are subequal. Color usually ranges from dark brown to black. The head is directed forward. The antennae are long and three segmented. Simple eyes, or ocelli, are present. The abdomen has one- or two-segmented urogomphi.

SIMILAR CALIFORNIA FAMILIES:

- sphaeritid beetles (Sphaeritidae)—body surface is shiny
- agrytid beetles (Agrytidae)—elytra with distinct grooves or striae

Rove Beetles (Staphylinidae)

Rove beetles are the largest family of beetles in California. Most species have short elytra exposing three or more of their abdominal segments. They occur in every kind of habitat from arctic tundra and alpine timberlines to tropical forests. Most species live in forest leaf litter and other decaying plant debris. Some species prefer to live in wet habitats, such as lakeshores and beaches. Dung and carrion are favored habitats for rove beetles that prey upon other insects. Other species live on trees and shrubs, preying upon aphids and bark beetles or feeding on pollen or fungi. Some species are specialists, living in the nests of mammals and birds, among ants and termites, or as parasites of insect pupae. A few of these specialists gain access to their hosts by mimicking their appearance and behavior. When excited, some harmless rove beetles often curl their abdomens over their backs, appearing as if they might sting. Many adults and some larvae produce noxious secretions that deter predators.

Little is known about the larvae of rove beetles. Most molt

Plate 29. The nocturnal Pictured Rove Beetle (*Thinopinus pictus*, Staphylinidae) (12 to 22 mm) may be found in great numbers throughout the year beneath seaweed along the length of the Pacific Coast, from Alaska southward to California. Both the adults and larvae hunt for small invertebrates, especially beach hoppers.

three times before becoming a pupa. They usually occupy the same habitats and have similar feeding habits as the adults. Some are parasites on the pupae of flies.

CALIFORNIA FAUNA: 1,230 species.

IDENTIFICATION: Most rove beetles (pl. 29) are black or brown and are usually small (1 to 10 mm), but a few are large (30 mm or more) and are brightly colored. The head and mouthparts are directed forward. The antennae are threadlike or with a club of four or more segments. The pronotum is usually broader than the head, with the borders often margined. The scutellum is visible. The elytra are short and often expose three or more abdominal segments. The tarsal formula is usually 5-5-5, but may be 4-5-5, 4-4-4, 5-4-4, or rarely with three or fewer segments.

The larvae are similar to those of the ground beetles (Carabidae). They are elongate, flattened, somewhat shaped like silverfish, and often covered with thick plates. The simple eyes are present or absent. The thorax has three well-defined segments, bearing four-segmented legs, each with a single, fixed claw. The 10-segmented abdomen usually ends in a two-segmented projection.

SIMILAR CALIFORNIA FAMILIES: Several other California beetle families have short elytra but may be distinguished from rove beetles by their overall appearance. Skiff beetles (Hydroscaphidae) are distinguishable from rove beetles by their last antennal segment, which is elongate.

Rain Beetles (Pleocomidae)

The family Pleocomidae contains the single genus *Pleocoma*, whose species are distributed from southern Washington southward to Baja California Norte, Mexico. Previously published records of rain beetles from Alaska and Utah have proved to be erroneous. California's rain beetles occur throughout the mountainous regions of the state, except in the deserts. Small, isolated populations also occur in the Sacramento Valley and the coastal plain of San Diego County. The known modern distribution of these apparently ancient beetles is restricted by the flightless females and is more or less correlated to areas of land that have never been subjected to glaciation or inundation by inland seas during the last two or three million years.

Rain beetles are large, robust, and shiny. The thick layer of hair covering the undersides is remarkably ineffective as insulation, especially for flying or rapidly crawling males who must maintain high body temperatures in cold, damp weather. Males can attain an internal temperature of 95 degrees F, although the mechanism by which they do this remains unclear. The thick pile probably functions to protect both sexes from abrasion as they burrow through the soil. Males and females dig with powerful, rakelike legs and a V-shaped scoop mounted on the front of the head. Males are fully winged and capable of flight, whereas the hind wings of the flightless females are reduced to small flaps of tissue. Females generally are much larger and more heavy bodied than males.

Lacking functional mouthparts, adult rain beetles are unable to feed. They must instead rely on fat stored in their bodies while they were root-feeding grubs. Because of their limited energy stores, adults are active for only a short time. On average, males of some rain beetles have only enough energy stored as fat to give them about two hours of air time and live only a few days. The more sedentary females require less energy and may live for months after fall and winter storms.

In most species of rain beetles, male activity is triggered by weather conditions that accompany sufficient amounts of fall or winter rainfall or snowmelt in late winter or early spring. Depending upon circumstances, males may take to the air at dawn or at dusk, or they may fly during evening showers. Others are encountered flying late in the morning on sunny days following a night of pouring rains, or during heavy snowmelt.

Males fly low over the ground, searching for calling females releasing pheromones from the entrances of their burrows. Amorous males are capable of tracking females over considerable distances, often through dense vegetation. Dozens of males may descend upon a single female, clambering over one another as they jockey for position to mate. Mating takes place on the ground or in the female's burrow. After mating, the males leave and may find another mate before coming to the end of their short lives. During their nuptial flights males are frequently attracted to lights or shiny pools of water. Females crawl back down their burrows and may wait up to several months for their eggs to mature. The female eventually lays 40 to 50 eggs in a spiral pattern at the end of the burrow as much as 3 m (10 ft) below the surface. The eggs hatch in about two months.

Upon hatching, the small grubs use their powerful legs

and jaws to tunnel deep in hard and compact soils to follow the root systems of their host plant. They feed upon roots of grasses, shrubs, and trees. In Oregon, the larvae of some rain beetles are considered pests when they attack the roots of strawberries, pears, apples, and cherries. Unlike most scarab beetles that have three instars, *Pleocoma* larvae molt seven or more times and may take up to 13 years before reaching maturity. Pupation occurs in a simple, elongate chamber.

Rain beetles have numerous enemies, below and above ground. Predatory fly maggots thought to be robber flies (Asilidae) attack both the larvae and pupae. Coyotes, foxes, skunks, raccoons, and owls feast upon the adults. Agile coyotes and foxes, as well as some other birds, even snatch the slow-flying males out of the air. At the height of the flight season it is not unusual to find the droppings of these predators filled with the indigestible bits of rain beetle legs and wing covers. One record describes a female infested internally by nematode worms.

CALIFORNIA FAUNA: 20 species and three subspecies in one genus.

IDENTIFICATION: Adult California rain beetles (pl. 30) are among the largest beetles in North America, resembling large June bugs in overall body shape. The males are usually reddish brown to black, whereas females are generally reddish brown. Their bodies are robust and broadly oval. The upper surface is shiny, whereas the underside is densely clothed in hairlike setae. The winged males (16.5 to 29 mm) are easily distinguished from the much larger, flightless females (19.5 to 44.5 mm) (pl. 31). The head is armed with a horn and bears antennae consisting of 11 segments tipped with four to eight fanlike segments that can be folded tightly into a club. The club is larger in males than females. Adults lack functional chewing mouthparts and do not feed. The prothorax is broad and unique among other scarab beetle families in that the cavities into which the front legs are in-

Plate 30. The Southern Rain Beetle (*Pleocoma australis,* Pleocomidae) (males 24 to 28 mm) is known from portions of the Transverse and Peninsular Ranges and flies during the first fall rains in October and November, usually during a light drizzle at dusk or just after dark. The larvae feed on the roots of canyon live oak *(Quercus chrysolepis).*

Plate 31. Female rain beetles, such as *Pleocoma badia badia* (32 to 43 mm), are flightless and much larger than their male counterparts.

serted are open toward the back. The scutellum is visible. The elytra are smooth, shining, or distinctly grooved and almost completely conceal the abdomen. The powerful forelegs are equipped with rakelike teeth. The tarsal formula is 5-5-5, with tarsal claws equal in size and simple. The abdomen has six segments visible from below.

The larvae are creamy white, C-shaped grubs. The tip of the abdomen may be darkened by waste material inside the body. The head is shiny yellowish or reddish brown. The antennae are three segmented, and the eyes are absent. The legs are four segmented. The 10-segmented abdomen does not have any projections.

SIMILAR CALIFORNIA FAMILIES: Their large size, hairy undersides, horned heads, 11-segmented antennae, and their fall and winter activity periods easily distinguish rain beetles.

Scarab Beetles, Dung Beetles, May Beetles, June Beetles, and Chafers (Scarabaeidae)

Scarabaeidae is one of the largest and most diverse families of beetles in California. These beetles have long been popular with collectors and naturalists because of their large size, beautiful colors, and interesting behaviors. Some species, especially in the tropics, are armed with horns or are clad in bright or metallic colors. The family includes one of the heaviest species in the world, the African Goliath beetle *(Goliathus),* weighing in at nearly three ounces. Two of the world's largest beetles are scarabs, the Elephant Beetle *(Megasoma elephas)* and the Hercules Beetle *(Dynastes hercules),* both from the New World. Some male Hercules Beetles measure 150 mm (6 in.) in length.

The dung-rolling Sacred Scarab *(Scarabaeus sacer),* depicted everywhere in ancient Egypt, from funerary art to hieroglyphs, is probably the most widely recognized species in the family. Carved scarabs bore religious inscriptions from the Book of the Dead or simply carried wishes for good luck, health, and life and were placed in the tombs of royalty from the earliest dynasties to ensure the success and well-being of the occupant in the afterlife. Heart scarabs were placed on or near the chest of the mummy and bore inscriptions admonishing the heart not to bear witness against its master on judgement day. Scarab mythology was so pervasive in ancient Egypt that carved scarabs were even worn as good luck amulets by Roman soldiers occupying Egypt for their presumed protective powers in battle. Today, many people wear scarabs as a symbol of good fortune, not unlike that of wearing a shamrock or rabbit's foot.

The life histories of the family are incredibly diverse. Eggs are laid in or near a suitable substrate, including dung, compost, or leaf litter, beneath the adult's host plant. The distinctive, C-shaped larvae, often called white grubs, molt twice before constructing a chamber and transforming into pupae. Some larvae (e.g., *Cotinis, Cremastocheilus, Euphoria*) construct pupal chambers from their own fecal material. Most species overwinter as larvae. Depending upon moisture and temperature conditions, adults emerge the following spring or summer and begin feeding on dung, compost, detritus, or roots. Some species are specialists, preferring to breed in the nests of ants, termites, birds, or rodents. Dung scarabs provision food for their young, but most other species simply lay their eggs and move on. Still, the biology and behavior of most California scarab beetles remain unknown.

Some California scarabs are diurnal or crepuscular, but most are nocturnal. In California, several genera (e.g., *Serica, Diplotaxis, Parathyce, Polyphylla, Cyclocephala*) are readily familiar to homeowners in residential areas and mountain communities because they are attracted to lights at night, sometimes in large numbers. The robust May beetles and

June bugs of the genus *Phyllophaga*, a common sight swarming about porch lights in the midwestern and eastern United States, are seldom encountered in California.

A few species are of minor concern as pests in California. The feeding activities of some larval *Aphodius* and *Ataenius* can damage lawns, especially in parks and golf courses. The larvae of the masked chafer *Cyclocephala* are also turf pests. Adult *Serica, Dichelonyx, Diplotaxis, Polyphylla,* and *Hoplia* may defoliate deciduous garden shrubs and orchard trees, as well as conifers in forests. Their larvae may be particularly destructive in forest nurseries, where they feed on the tender roots of seedling trees.

CALIFORNIA FAUNA: 288 species in 46 genera.

IDENTIFICATION: California scarab beetles (pl. 32; see also pls. 19, 21, 49) vary considerably in shape and are oval, long, square, cylindrical, or slightly flattened and range in length up to 30 mm or more. They are black, brown, yellowish brown, and occasionally metallic or scaled with blotched or striped patterns. The head is directed weakly downward. The eight- to 10-segmented antennae are tipped with three to seven fanlike segments that fold tightly into a flat, lopsided club. The antennal club appears velvety in the dung scarabs but is nearly bare in all other species. The pronotum is variable, with or without horns and tubercles, but the sides are always distinctly margined. The scutellum is hidden in the dung beetle genus *Onthophagus* but visible in all others. The elytra are slightly rounded or parallel sided, with the surface smooth, distinctly pitted, grooved, or covered with scales, and completely conceal the abdomen only in the dung scarabs. The tarsal formula is 5-5-5, with the claws equal in size or not and simple or toothed. The abdomen has six segments visible below.

The whitish, yellowish, or creamy white larvae, or grubs, are C shaped. In some dung scarabs (e.g., *Canthon, Liatongus,* and *Onthophagus*) the larvae appear humpbacked because of the enlargement of the thoracic segments. The head is distinct

Plate 32. The Ten-lined June Beetle (*Polyphylla decemlineata,* Scarabaeidae) (18 to 31 mm) is widely distributed throughout western North America. It is found in all areas of California except the Mojave and Colorado Deserts. The larvae feed on the roots of various plants, whereas the adults are recorded to feed on the needles of ponderosa pine *(Pinus ponderosa).* Adults are commonly attracted to lights.

and darker than the rest of the body and has a pair of four-segmented antennae. The simple eyes are either present, faintly indicated by pigmented spots, or absent. The legs are two segmented in some dung scarabs (e.g., *Canthon, Liatongus,* and *Onthophagus*) and four segmented in all other groups. The 10-segmented abdomen lacks projections.

SIMILAR CALIFORNIA FAMILIES:

- stag beetles (Lucanidae)—antennae elbowed, with club segments unable to form a compressed club
- false stag beetles (Diphyllostomatidae)—antennae with club segments unable to form a compressed club; mandibles usually protruding; abdomen with seven segments visible below
- enigmatic scarab beetles (Glaresidae)—abdomen with five segments visible below; first segment of antennal club cup shaped to receive remaining segments

- hide beetles (Trogidae)—abdomen with five segments visible below
- rain beetles (Pleocomidae)—antennae 11-segmented; mandibles absent
- earth-boring scarabs (Geotrupidae)—antennae 11-segmented
- sand-loving scarab beetles (Ochodaeidae)—mandibles exposed; spurs on ends of middle tibiae appear feathery or finely notched
- scavenger scarab beetles (Hybosoridae)—first segment of antennal club cup shaped to receive remaining segments
- bumblebee scarabs (Glaphyridae)—elytra short with distinctly separated tips; abdominal segments exposed

Jewel Beetles, or Metallic Wood-boring Beetles (Buprestidae)

Metallic wood-boring beetles, or jewel beetles, are named for their beautiful and iridescent colors. Many species are brightly marked with yellow, orange, or red bands and spots. Their streamlined, bullet-shaped bodies resemble those of click beetles (Elateridae); however, their rigid bodies, saw-toothed antennae, and metallic colors easily distinguish them from the click beetles. Most species are somewhat flattened, as indicated by their elliptical or oval emergence holes left behind in trunks and branches. Although jewel beetles are among the most destructive of borers in managed timber regions, they are an important link in the recycling of dead trees and downed wood.

Adult jewel beetles feed on foliage, pollen, and nectar. Active on hot, sunny days, they readily fly when disturbed. Most species are strong flyers, and some make a loud buzzing noise when they are airborne. They are often seen running rapidly over trees, probing the bark and wood with their ovipositors in preparation for laying their eggs.

Although a few species attack healthy trees, most jewel beetles prefer to lay their eggs on trees or shrubs weakened by drought, injury, or infestations by other insects. Recently killed trees, especially those felled by logging operations, are particularly attractive. Females lay their eggs directly on the trunks and branches or in crevices in the bark or wood.

The legless larvae are often shaped like horseshoe nails by virtue of their broad, flat thoracic segments, suggesting the name "flatheaded" borers. Many flatheaded borers mine the sapwood of branches, trunks, and roots, whereas others bore extensively into the heartwood. Some species work both. Their tunneling activities can hasten the death of already weakened trees. The larvae of *Agrilus* produce knotty swellings of living plant tissues known as galls. Girdlers construct spiral galleries around small stems, killing the terminal end of the branch. Still other species (*Brachys* and *Taphrocerus*) are stem and leaf miners of herbaceous and woody plants. A few flatheaded borers attack cones (*Chrysophana*) or seasoned wood (*Buprestis*).

The larval galleries are broad, flat, and form long linear or meandering tracts beneath the bark or in the heartwood. The galleries are usually tightly packed with dust and frass. This material is often arranged in finely reticulate ridges that resemble fingerprints. Most species overwinter as larvae, but a few species pupate in fall and overwinter as adults. The time of development from egg to adult may take one or more years. Individuals of some species may take longer under extraordinary circumstances.

CALIFORNIA FAUNA: Approximately 264 species in 33 genera.

IDENTIFICATION: California jewel beetles, or wood-boring beetles (pl. 33), are elongate, flattened, or cylindrical beetles ranging in length up to 31 mm. They are usually metallic or black with orange, red, or yellow markings. The head is tightly

Plate 33. The Golden Buprestid (*Cypriacis aurulenta,* Buprestidae) (12 to 22 mm) is most common in the pine forests of the Sierra Nevada and the Transverse and Peninsular Ranges. Females lay their eggs in and around fire scars and wounds on conifers, especially ponderosa pine *(Pinus ponderosa),* Douglas-fir *(Pseudotsuga menziesii),* fir *(Abies),* and spruce *(Picea).*

tucked inside the slightly broader prothorax. The mouthparts are directed downward, and the serrate antennae consist of 11 segments. The scutellum may or not be visible. The elytra are smooth, ribbed, or sculptured and usually almost completely conceal the abdomen. The tarsal formula is 5-5-5, with each tarsal claw equal in size and simple, lobed, or notched. The abdomen has five segments visible from below, the first two of which are distinctly fused together.

The creamy white, white, or yellowish larvae are long, slender, and legless. Their small eyeless head bears three-segmented antennae and can be retracted within the much broader thoracic segments. The thoracic segments have a distinct, V-shaped groove on the back. The slender, 10-segmented abdomen is sharply narrowed behind the thorax and is tipped by two fleshy lobes, or with a pair of short, thick projections or toothed forks.

SIMILAR CALIFORNIA FAMILIES:

- false click beetles (Eucnemidae)—never metallic; body distinctly flexible between prothorax and elytra
- click beetles (Elateridae)—rarely metallic; body distinctly flexible between prothorax and elytra
- lizard beetles (Languriidae)—antennae clubbed

Click Beetles (Elateridae)

Click beetles are found throughout California, reaching their greatest diversity in the wetter, montane regions of the state, particularly in forests and meadows. They are easily recognized by the large, loosely joined forebody. Click beetles are commonly found on vegetation or under bark and are often attracted to lights. Although some are active during the day, most species are nocturnal. Many species feed on rotting fruit, flowers, nectar, pollen, fungi, and sapping wounds on shrubs and trees. Predatory species attack small invertebrates, especially wood-boring insects and plant hoppers.

The name "click beetle" is derived from the distinctly audible sound made by a special mechanism on the underside of the body between the prothorax and mesothorax (see chapter 2). The sudden noise is probably intended to startle predators. Stranded on their backs, click beetles can also use this mechanism to right themselves by flipping up into the air, although several attempts may be required before landing on their feet. Smaller species can propel themselves up to 25 cm (10 in.) into the air, whereas larger species (e.g., *Alaus* and *Chalcolepidius*), flip only just a few centimeters.

The tough, wiry larvae are called wireworms and resemble slender mealworms sold in pet stores and bait shops. Wireworms are found in soil, rich humus, or decaying plant materials, especially rotten wood, where they are primarily predators. The larvae of wood-dwelling species prey upon small invertebrates, whereas the larvae of other species scavenge

Plate 34. The western Eyed Click Beetle (*Alaus melanops,* Elateridae) (20 to 35 mm) is found in the Sierra Nevada, the Cascade Range, and the Transverse and Peninsular Ranges. The larvae are found beneath the bark of stumps and standing or fallen logs throughout the year and attack various wood-boring insect larvae. Adults are attracted to lights and reach their peak of activity from May through July.

fungi. Soil dwellers feed on invertebrates, roots, or both. Pest species damage seeds and roots of a variety of crops and garden plants. The larvae undergo three to five molts and may take up to three years to mature, depending upon the availability of food and moisture. In rare incidences the larvae may develop stunted wings prematurely. This condition is known as *prothetely* and is apparently induced by unfavorable environmental conditions. It occurs most often in wireworms reared in the laboratory. Both adults and larvae overwinter in the ground, under bark, or in rotten wood.

CALIFORNIA FAUNA: Approximately 300 species.

IDENTIFICATION: Click beetles (pl. 34) are easily distinguished from other beetle families by their long, somewhat flattened bodies and the large, freely moving prothorax, which usually has backward-pointing projections on the back corners. They are brown or black in color, but a few species bear distinct markings or have a slightly metallic surface *(Chalcolepidius)*. They range in length up to 35 mm. The head is directed slightly downward, with the mandibles exposed (e.g., *Aplastus, Euthysanius, Octinodes*) or not. The 11-, or rarely 12-segmented antennae are usually saw toothed or feathery (e.g., *Aplastus, Euthysanius, Octinodes*) and attached close to the eyes. The scutellum is visible. The elytra are smooth, ribbed, occasionally hairy or scaly, and nearly always completely conceal the abdomen. The tarsal formula is 5-5-5, with the claws equal in size and simple, toothed, or comblike. The abdomen has five segments visible below.

The larvae are elongate, slender, and cylindrical, with most of the abdominal segments similar in length. The color varies from pale white to dark yellowish brown. The head is wedge shaped and directed forward. The antennae are three segmented. Simple eyes range from zero to six on either side of the head. The legs are long and five segmented. The 10-segmented abdomen terminates simply or ends in spines, bumps, or bumplike projections.

SIMILAR CALIFORNIA FAMILIES:

- soft-bodied plant beetles (Dascillidae)—forebody not moveable
- false click beetles (Eucnemidae)—forebody not moveable; body often widest at pronotum; antennae not attached close to eyes
- throscid beetles (Throscidae)—antennae clubbed, rarely saw toothed
- metallic wood-boring beetles (Buprestidae)—usually metallic underneath
- lizard beetles (Languriidae)—antennae clubbed
- some false darkling beetles (Melandryidae)—tarsi 5-5-4

Phengodid Beetles (Phengodidae)

Phengodidae is a small New World family of soft-bodied beetles closely related to fireflies (Lampyridae) and soldier beetles (Cantharidae). The family is mostly tropical in distribution, with six genera occurring in North America. Phengodids are particularly interesting because of the larviform condition of adult females and the brilliant bioluminescence of the eggs, larvae, and adult females. Sometimes called glowworms, the dorsal surface of both thoracic and abdominal segments possesses luminous bands. The intensity of these lights may vary, but they do not flash on and off as in some fireflies. It has been suggested that the males are first attracted from a distance to the pheromones released by a calling female. Once in the vicinity, the males can zero in on her lights.

Most phengodid larvae are bioluminescent, with spots of green, orange, or rarely red light. The larvae of the South American genus *Phrixothrix* are unusual in producing two colors of light. The head glows a fiery red and is followed by a series of pale greenish yellow lights on the abdomen.

Phengodid larvae are predaceous upon millipedes, insects, and possibly other invertebrates, whereas adult males and larviform females are not known to feed at all. Larvae live in leaf litter under forest trees or hide beneath bark or boards on the ground.

CALIFORNIA FAUNA: Nine species in five genera.

IDENTIFICATION: Adult males (pl. 35) are soft bodied, elongate, and flattened in overall body shape. They range in length from 12 to 23 mm. The head is directed forward and fitted with a pair of short sickle-shaped mandibles. The 12-segmented antennae are feathery. The pronotum is flat, quadrate, and sharply margined on each side. The scutellum is visible. The elytra of the male are soft and do not conceal the last three or more segments of the abdomen. The tarsal formula is 5-5-5, with all claws equal in size and simple. The abdomen has seven segments visible below.

The distinctly banded larvae are elongate, slightly flattened, straight-bodied insects that are somewhat tapered at both ends. Mature female larvae may reach 40 to 65 mm in length. The mouthparts project forward and are born on a retractable head. The head also bears three-segmented antennae, and simple eyes are absent. Each four-segmented leg bears a single, sharp claw. The last abdominal segment is without projections. Adult females are similar in external appearance to the larvae but can be readily distinguished by the presence of compound eyes and a genital opening on the underside of the ninth abdominal segment.

SIMILAR CALIFORNIA FAMILIES: The protruding mandibles, distinct form and color, and feathery antennae of the males distinguish phengodids from other California beetle families.

Plate 35. The Western Banded Glowworm (*Zarhipis integripennis*, Phengodidae) (males 12 to 23 mm) is found in all regions below 6,000 feet except the Great Basin Desert and much of the Colorado Desert. The males are frequently attracted to lights in spring, and the larvae feed on millipedes.

Fireflies and Glowworms (Lampyridae)

Fireflies and lightning bugs are neither flies nor true bugs, but flat, soft-bodied beetles. Some adult females look more like grubs than beetles and are called glowworms. Like the phengodids (Phengodidae), the larvae of these females undergo a modified pupal stage before becoming sexually mature. Larviform females (pl. 9) are distinguished from larvae externally by their compound eyes.

Of the more than 140 species of fireflies and glowworms found in North America, most occur in more tropical and subtropical climates. People from eastern or southern United States often remark on the lack of lightning bugs, or fireflies, in California. Eighteen species, however, are represented in the state, but not all of them bioluminesce and none of them produce flashing lights while flying to communicate.

The light-producing organs of fireflies, if present, are whitish or yellowish in color and located on the underside of the abdomen. Even the eggs, larvae, and pupae of some species glow, but the purpose for this remains unclear. Male fireflies have very large eyes adapted for locating light-producing females. The color of the light varies among species from light green to orange and may be determined by temperature and humidity. The light-producing organs are located underneath the fifth abdominal segment in females and the fifth and sixth segment of males. These organs are supplied with air by numerous breathing tubes, or tracheae. By regulating the oxygen supply to these organs, fireflies can control the brightness and frequency of their light.

Bioluminescence in fireflies is virtually 100 percent efficient, with almost all the energy that goes into the system given off as light. An incandescent light bulb is not nearly as efficient, with 90 percent of the electrical energy lost as heat. In fact, the light-producing organs of one firefly produce

Plate 36. The California Glowworm (*Ellychnia californica,* Lampyridae) (9.5 to 16 mm) is not bioluminescent. It is found throughout California where adults are found on flowers or low on grassy vegetation, particularly in moist habitats. The larvae are probably active at night and prey on snails.

1/80,000th of the heat produced by a candle flame of the same brightness.

Nocturnal, light-producing species spend their days hiding beneath bark or in leaf litter or resting on leaves. Fireflies with weak or no light-producing organs, such as *Ellychnia,* are active during the day and are generally found on flowers or on streamside vegetation. The feeding habits of adult fireflies are poorly known, but many species do not feed at all.

The predatory larvae attack snails and slugs, earthworms, and small insects such as cutworms. Chemicals are pumped through hollow mandibles to paralyze and liquefy prey, and the tissues are then swallowed.

CALIFORNIA FAUNA: 18 species in seven genera.

IDENTIFICATION: Male California fireflies and glowworms (pl. 36) are elongate, flattened, and soft-bodied beetles ranging in length from 4.2 to 16 mm. They are usually pale brown, black and reddish brown, or black with some red or pink markings. The head is concealed by the leading edge of the pronotum. The mouthparts are directed downward, and the eyes are relatively large. The antennae are 10- (as in *Microphotus*) or 11-segmented and are usually threadlike, saw toothed, or comb-like. The pronotum is triangular and flattened and distinctly margined at the sides. The scutellum is visible. The elytral margins are nearly parallel, widest at the middle, and almost or completely conceal the abdomen. The surface may have a few faint ribs running lengthwise. The tarsal formula is 5-5-5, with the next-to-last segment of each foot heart shaped. The claws are equal in size and simple. The abdomen has eight (males) or seven (females) segments visible below.

The larvae are spindle shaped or may resemble somewhat flattened pillbugs, with distinct armored segments. They vary in color from dark brown to black, or pink. The head is retracted into the thorax, with a pair of three-segmented antennae and two pairs of simple eyes. The legs are four segmented. The 10-segmented abdomen lacks projections at the tip.

SIMILAR CALIFORNIA FAMILIES:

- net-winged beetles (Lycidae)—elytral ribs connected by distinct but less-conspicuous cross veins
- phengodid beetles (Phengodidae)—head exposed; abdomen slightly exposed; sicklelike mandibles distinct; antennae feathery
- soldier beetles (Cantharidae)—head exposed

Soldier Beetles (Cantharidae)

Soldier beetles are frequently encountered on flowers and foliage. In the drier regions of California they prefer living on streamside plants. The short-lived adults are usually active during spring and summer days. They feed mostly on liquids high in nutrients obtained from nectar or insect prey. Carnivorous species, especially those that feed on aphids (Aphididae) and their relatives, are considered to be of some use as biological control agents. The larvae, pupae, and adults produce defensive secretions from their abdominal glands. The contrasting red and black, bluish black, or gray colors of the adult beetles warn potential predators of their bad taste. Distasteful species may serve as models for other beetles to mimic their appearance in order to discourage predators. When disturbed, some soldier beetles withdraw their legs and drop to the ground.

The larvae may live from one to three years; live under bark or in damp areas beneath stones, logs, or other objects on the ground; and are carnivorous or omnivorous. Carnivorous species feed on caterpillars, maggots, and grasshopper eggs. The larvae of some plant-feeding species attack grasses, potatoes, and celery.

The adults of *Cantharis, Chauliognathus,* and *Podabrus* suffer apparently lethal infections of a zygomycetous fungal pathogen *(Eryniopsis)* known to attack insects. Dead beetles are literally found in a death grip, with their mandibles embedded in the tissue of a leaf and their bodies twisted up between the extended wings.

CALIFORNIA FAUNA: 159 species in 10 genera.

IDENTIFICATION: Adult soldier beetles (pl. 37) are long, soft-bodied beetles that resemble fireflies and are usually dark brown to black, sometimes with yellow, orange, or red markings. The head is directed forward or sometimes downward. The 11-segmented antennae are usually threadlike. The pronotum is usually broader than the head and is typically wider than long. The scutellum is visible. The elytra are smooth (sometimes appearing velvety), soft, flexible, and parallel sided and nearly or completely conceal the abdomen. The tarsal formula is 5-5-5, with the fourth segment appearing heart shaped. The claws are equal in size and are simple, toothed, or lobed. The abdomen has seven (females

Plate 37. The Brown Leatherwing Beetle (*Cantharis consors*, Cantharidae) (12 to 20 mm) is found from British Columbia southward to Baja California. Adults are common visitors to porch lights in late spring and early summer and emit a musty odor when handled or crushed. The adults feed on the Citrus Mealybug *(Plancoccus citri)*, whereas the larvae probably live among plant litter and prey on other small insects.

and some males) or eight (most males) segments visible from below.

The larvae are silverfish shaped and clothed with a covering of fine hairs that gives them a velvety appearance. The head is strongly directed forward, usually with a large, simple eye on each side. The antennae are three segmented. The legs are five segmented. The 10-segmented abdomen has paired glands on the first nine segments that produce defensive secretions and does not end in any projections.

SIMILAR CALIFORNIA FAMILIES:

- lightning bugs and glowworms (Lampyridae)—head covered by pronotum; abdomen sometimes with light-producing organs
- net-winged beetles (Lycidae)—elytra with raised network of veins

- phengodid beetles (Phengodidae)—antennae feathery; mandibles sickle shaped
- blister beetles (Meloidae)—bodies more cylindrical; head with neck; tarsi 5-5-4
- false blister beetles (Oedemeridae)—prothorax without distinct margins, tarsi 5-5-4
- some longhorn beetles (Cerambycidae)—tarsi apparently 4-4-4

Skin Beetles (Dermestidae)

Skin beetles are primarily scavengers that feed on materials high in protein, including fur, feathers, and carcasses, as well as pollen and nectar. A few species feed on cork, seeds, grains, and other cereal products. The adults and larvae of some species are important economic pests and cause millions of dollars worth of damage annually.

Household and museum pests, such as *Attagenus* and *Anthrenus,* resemble small lady beetles mottled with black, brown, tan, and white hairs or scales. Adults feed largely on pollen and nectar but may enter homes and other buildings in spring and summer to lay their eggs on animal products. The larvae develop in dark, undisturbed places and are solely responsible for damaging woolen materials, carpets, silk products, dried meats, and museum specimens, including collections of insects (see chapter 5).

The clusters of bristly hairs found on the bodies of the larvae serve as an irritating deterrent to predators, such as mammals, reptiles, and birds, and are also known to cause human allergies. They are found in house dust and have recently been linked to asthma. These same hairs are thought to entangle the mouthparts of ants and other small arthropod predators. Clusters of hairs, located on the membranes between some of the upper abdominal plates, can be raised and spread into a protective fan as a defense against predators.

Plate 38. The Common Carrion Dermestid (*Dermestes marmoratus,* Dermestidae) (10 to 13 mm) is associated with the latter stages of decomposition of a carcass, where both the adults and larvae feed on the dried flesh. Adults move with quick, jerky movements.

With the increased use of synthetics in furniture making and carpets, skin beetles are not as common in homes as they used to be. They occasionally, however, attack synthetic fibers stained with sweat, urine, food, or drink. Household reinfestations of skin beetles may be the result of undetected natural reservoirs, such as bird, mammal, or paper wasp nests. Spider webs, window sills, and light fixtures containing dead insects are also breeding grounds for some species of skin beetles. Even some insulation materials wrapped around electrical lines hidden in walls are potential breeding sites for these pests.

CALIFORNIA FAUNA: Approximately 25 species in 11 genera.

IDENTIFICATION: Skin beetles (pl. 38; see also pl. 16) are compact, nearly oval or round beetles, ranging in length up to 13 mm. The head is directed downward and is capable of being retracted into the prothorax. The five- to 11-segmented antennae are clubbed and attached in front of the eyes. A simple eye is usually present between the compound eyes. The pronotum is broader than long and narrowed toward the head. The side margins of the prothorax are distinct, or at least finely ridged toward the back. The scutellum is usually visible. The elytra are smooth or distinctly ribbed, clothed with hair or scales, and completely conceal the abdomen. The tarsal formula is 5-5-5, with claws equal in size and simple. The abdomen usually has five segments visible from below.

The larvae are long or almost oval and nearly cylindrical, or oval and flattened. The dark body is covered with clumps of short or long bristly hairs. The head is directed downward, with three-segmented antennae. Up to six pairs of simple eyes are present. The legs are four segmented. The abdomen is nine or 10 segmented, with projections present dorsally on the ninth segment in *Dermestes* but absent in all other genera.

SIMILAR CALIFORNIA FAMILIES: Skin beetles are unique in that most of them have a simple eye located between the compound eyes.

- death watch beetles (Anobiidae)—antennae longer, and club, if present, lopsided
- pill beetles (Byrrhidae)—antennal club, if present, formed gradually
- plate-thigh beetles (Eucinetidae)—antennae threadlike
- wounded-tree beetles (Nosodendridae)—legs with tibiae expanded

Powder-post Beetles, Twig Borers, or Branch Borers (Bostrichidae)

Powder-post beetles, twig borers, or branch borers live either in living trees or dead wood. The tunneling activity of both the adults and larvae often reduces the wood to powder. In fact, the name "powder-post" beetle is derived from the way one group of these beetles (formerly Lyctidae) leaves behind a tube or post of fine, powdery frass inside tunneled wood. Other larvae leave galleries tightly packed with coarser dust and frass mixed with wood fragments. The closely placed exit holes of some species of these beetles have suggested another common name, the shot-hole borers.

Powder-post beetles, twig borers, or branch borers are especially fond of dead branches and fire-killed hardwoods. Some species prefer old wood, whereas others attack cut and seasoned wood. The tunneling activities of these larvae are particularly damaging to old dwellings and furniture. Others mine through living limbs of weakened cultivated trees or tunnel through green shoots of living plants, whereas some tropical species attack felled timber and bamboo. Because of their tendency to bore into wood products, powder-post beetles are widely distributed around the world through commerce. Some species cause considerable damage to stored products, especially dried roots and grains, whereas others prefer to feed on woody fungi.

The Giant Palm Borer *(Dinapate wrightii)* (30 to 52 mm) is the largest bostrichid in the world. It is found in the desert oases of Baja California and the Colorado Desert where California fan palms *(Washingtonia filifera)* occur. To hear both adults and larvae chewing inside palm trunks from several feet away is not unusual. Once considered rare, they are now pests of planted palms in Arizona and California.

CALIFORNIA FAUNA: 27 species in 18 genera.

IDENTIFICATION: The Bostrichidae (pl. 39) are elongate and cylindrical or somewhat flattened in overall body shape and are usually black, dark brown, or reddish brown. They range in length from 2 to 23 mm (rarely to 52 mm). The head is usually directed downward and is not generally visible from above and may or may not be covered by the pronotum. The eight- to 11-segmented antennae are straight and tipped with a two- to four-segmented club. The pronotum is somewhat square, and the rounded margins are with or without

Plate 39. The Giant Palm Borer (*Dinapate wrightii*, Bostrichidae) (30 to 52 mm) is found in the desert oases of Baja California and the Colorado Desert where California fan palms *(Washingtonia filifera)* occur. It is not unusual to hear both adults and larvae chewing inside palm trunks from several feet away. Once considered rare, they are now pests of planted palms in Arizona and California.

fine teeth. The front of the pronotum may be rough and rasp-like and is sometimes armed with small horns. The scutellum is visible. The elytra are parallel sided and variously modified with lines of punctures, ridges, or apical spines and completely conceal the abdomen. The elytral surface is coarsely or finely punctate. The apices of the elytra may or may not slope sharply downward to give the beetle an abruptly cut-off look. The tarsal formula is 5-5-5, with each claw equal in size and toothed. The abdomen has five segments visible from below.

The cream or dull white larvae are C shaped with a dark head that can be retracted or extended forward. The antennae are three segmented, and simple eyes or eye spots are present. The thoracic segments are enlarged, whereas the smaller abdominal segments are approximately equal in length. The legs are three or four segmented. The tip of the abdomen is without projections.

SIMILAR CALIFORNIA FAMILIES:

- Rugose Stag Beetle (Lucanidae)—antennae lamellate; head directed forward; male horned
- some death watch beetles (Anobiidae)—antennae saw toothed or fan shaped
- cylindrical bark beetles (Colydiidae)—mandibles concealed
- bark or ambrosia beetles (Curculionidae)—antennal club compact

Checkered Beetles (Cleridae)

Checkered beetles are long, slender, slightly hairy, and bug-eyed beetles that are sometimes brightly marked with distinct patterns. Some strikingly colored species that are active during the day are thought to mimic wasps, moths, flies, and other beetles. These species are boldly marked with black and red or yellow patterns or uniformly shiny green or blue. Checkered beetles are found in a wide variety of habitats, from flowers and beetle-infested trees to dried carcasses and stored foods.

Both larvae and adults prey on other insects. The larvae of checkered beetles are associated with dead wood where they prey on the immature stages of bark beetles (Curculionidae), longhorn beetles (Cerambycidae), and metallic wood-boring beetles (Buprestidae), whereas the adults (e.g., *Enoclerus*) run up and down branches hunting adult beetle prey and searching for mates and egg-laying sites. The drab and nocturnal *Cymatodera* feed and lay their eggs at night, hiding during the day beneath bark or other debris. The larvae of *Trichodes* develop in the nests of bees and wasps, whereas those of *Aulicus* attack the underground egg pods of grasshoppers. The predatory *Necrobia* infests stored food products (see chapter 5).

CALIFORNIA FAUNA: 76 species and seven subspecies in 22 genera.

Plate 40. The Ornate Checkered Beetle, or Common Checkered Beetle, (*Trichodes ornatus,* Cleridae) (5 to 15 mm) is found on a wide variety of flowers. It feeds on pollen and preys on small, flower-visiting beetles. Mating takes place on flowers, after which the female may consume the male. The larvae are predators in the nests of leaf-cutter bees and wasps.

IDENTIFICATION: California checkered beetles (pl. 40; see also pl. 15) are long, slender beetles typically clothed in moderately long, erect hairs and range in length from 2 to 20 mm. The head is directed downward and is usually as wide or wider than the prothorax. The antennae have eight to 11 segments and are usually abruptly clubbed but are sometimes threadlike, saw toothed, or comblike. The pronotum is longer than wide and narrower than the base of the elytra. The scutellum is visible. The elytra almost or completely conceal the abdomen. The tarsal formula is 5-5-5, with all claws equal in size and simple or toothed. The abdomen has five or six segments visible from below.

The larvae are variously colored and mottled with black or shades of brown and are long, slender, and straight bodied. Some are uniform in width, and others are markedly thicker at the middle of the abdomen. The distinct head is directed forward and bears three-segmented antennae and zero to five pairs of simple eyes. The four-segmented legs are widely separated. The 10-segmented abdomen may have fleshy folds or wrinkles. Branched or unbranched projections at the tip of the abdomen are present or absent.

SIMILAR CALIFORNIA FAMILIES:

- antlike flower beetles (Anthicidae)—head with neck
- antlike leaf beetles (Aderidae)—head with neck
- soft-winged flower beetles (Melyridae, Dasytinae)—pronotum margined; antennae usually saw toothed
- telephone-pole beetles (Micromalthidae, not established in California)—antennae beadlike
- narrow-waisted bark beetles (*Elacatis* spp., Salpingidae)—head directed forward
- soldier beetles (Cantharidae)—antennae long, threadlike

Lady Beetles (Coccinellidae)

Also known as ladybird beetles or ladybugs, lady beetles are among the most recognized and beloved insects in the world. These beetles, with their bright contrasting red and black coloration, have attracted attention for centuries in spite of the fact that only a fraction of the world's 4,500 species are so colored. In the Middle Ages, certain species of lady beetles were dedicated to the Virgin Mary and named beetles of Our Lady. A considerable body of folklore is associated with these beetles. They are widely equated with good luck and are often associated with happy events, such as weddings. Consequently, in many parts of the world, the harming of one of these beetles is thought to bring bad luck.

Lady beetles are surprisingly diverse in their habits. Many species, particularly in the tropics, are plant feeders and are considered to be pests. A few feed on molds and fungi, but the vast majority of species are predaceous on aphids, scale insects and mealy bugs, and even mites. As such, they are considered among the most beneficial of insects and are commonly used as biological control agents (see chapter 1). Exotic species are introduced to control various target pests. By 1985 more than 100 species of exotic lady beetles had been introduced into California. Aphid-feeding species are routinely collected in large numbers and released to control aphids. These efforts have sometimes displaced native lady beetles, increased predation on nontarget insects, or created a public nuisance by introducing species that congregate inside homes and on buildings.

The bright colors of many lady beetles advertise the presence of repellent chemicals in their bodies. The yellowish fluid released from their knees contains toxic alkaloids that render these beetles distasteful to predators. Other colorful species that do not possess these chemicals mimic the distasteful species and thus falsely advertise themselves as unpalatable to predators. The color patterns on many larvae apparently serve the same functions.

CALIFORNIA FAUNA: Approximately 180 species.

IDENTIFICATION: California lady beetles (pl. 41) are typically oval to long and oval or round in outline and weakly to strongly convex in profile. The antennae are seven to 11 segmented and are more or less clubbed. The head is deeply inserted into the prothorax and usually not visible from above. The scutellum is visible. The elytra are smooth and cover the abdomen completely. The abdomen has five or six segments visible from underneath; a seventh segment is rarely visible. The first abdominal segment immediately behind the attachment of the hind legs usually has at least one distinct line. The tarsal formula appears 3-3-3 but is actually 4-4-4. Tarsal claws are simple or toothed.

The larvae (pl. 6) are elongate and frequently bear spines, tubercles, or wrinkles. They are frequently marked with

Plate 41. Convergent Ladybirds (*Hippodamia convergens,* Coccinellidae) (4.2 to 7.3 mm) migrate every March and April from overwintering sites in the Sierra Nevada to the Great Central Valley. Larval activity reaches its peak in April, followed by pupation in May. As summer temperatures warm and food supplies dwindle, the adults take wing and fly up into the cooler, mountain canyons to feed and overwinter.

white, yellow, orange, or red against a grayish or black ground color. The head is directed forward and bears up to three pairs of simple eyes. The antennae are one to three segmented. The legs are long, particularly in active predaceous species, and four segmented. The 10-segmented abdomen lacks projections.

SIMILAR CALIFORNIA FAMILIES:

- round fungus beetles (Leiodidae)—antennal club distinct; elytra often grooved
- marsh beetles (Scirtidae)—antennae threadlike or saw toothed
- shining flower beetles (Phalacridae)—antennal club distinct; tarsi appear 4-4-4, but are 5-5-5; no distinct markings on the elytra

- pleasing fungus beetles (Erotylidae)—antennal club distinct
- handsome fungus beetles (Endomychidae)—front angles of pronotum distinctly pointed forward
- minute fungus beetles (Corylophidae)—antennal club distinct
- leaf beetles (Chrysomelidae, Chrysomelinae)—tarsi appear 4-4-4 but are actually 5-5-5

Darkling Beetles (Tenebrionidae)

Darkling beetles are among the most conspicuous beetles in California. The head-standing stink beetles *(Eleodes)* are among some of the best known members of the family. They are extremely variable in shape and size. Most species are black, hard-bodied, and long-lived beetles known to live 10 years or more. They are found in nearly all habitats, except water, and are particularly abundant in the drier regions. To conserve water and regulate body temperatures, species living in deserts or along the coast often have sealed cavities beneath their wing covers to reduce the amount of water lost as they breathe. They also take shelter inside rodent burrows, among leaf litter, or beneath stones, bark, or logs.

Many California species are sand dune obligates, requiring fine loose sand of beaches and deserts in which to burrow, feed, and breed. Steady plodders or vigorous runners, these dune dwellers often leave long and familiar tracks winding over the fine sand.

At dusk or in the evening, darkling beetles are usually found crawling on downed logs, up-standing tree trunks, or over the ground. Species associated with trees are found either under bark or in galleries and tunnels carved out by wood-boring insects. Some prefer living in rotten trunks, feeding on wood riddled with fungi. Others prefer to feed on the fruiting bodies of the fungi themselves. Still other species graze upon the surfaces of lichens, algae, and mosses growing on the surfaces of bark and rocks.

Soil-dwelling larvae and adults are sometimes extremely abundant in drier regions, where they feed on the seeds, roots, and detritus of many kinds of plants. Chaparral and desert dwellers sometimes use the seed-strewn mounds of harvester ants (*Messor* and *Pogonomyrmex*) like a buffet, scavenging bits of plant detritus.

The long, cylindrical larvae are sometimes called false wireworms because of their resemblance to larval click beetles (Elateridae). The front pair of legs of soil- and sand-dwelling larvae are often much thicker and more heavily armored than the remaining legs. The immature stages of the mealworms *Tenebrio* and *Zophobas*, sold in pet shops and bait stores, are familiar examples of darkling beetle larvae (see chapter 5). Both the larvae and adults are seldom of any negative economic importance.

CALIFORNIA FAUNA: Approximately 250 species.

IDENTIFICATION: California darkling beetles (pl. 42) are usually black or dark brown, ranging in length up to 30 mm or more. They are extremely variable in shape, ranging from elongate and somewhat cylindrical to oval or round and strongly humpbacked or hemispherical. The head is usually directed forward or occasionally slightly downward. The usually 11-segmented antennae are beadlike or gradually clubbed and are attached under the frons, which forms an expanded shield with the canthus and extends well beyond the head. The eyes are usually notched. The pronotum is either broadly rounded or hatchet shaped and is usually distinctly margined on the sides. The scutellum is visible. The elytra completely conceal the abdomen and are smooth, pitted, bumpy, grooved, or ridged. Species with elytra partially or completely fused have their flight wings reduced or absent. The legs are variable, from stout and specialized for digging, to somewhat slender for rapid movement. The tarsal formula is 5-5-4, rarely 4-4-4,

Plate 42. When disturbed, many species of *Eleodes* (Tenebrionidae) first stand on their heads and then discharge noxious chemical compounds. This defensive behavior has earned them the name acrobat, clown, or stink beetles. The chemical compounds act as repulsive agents and are laced with hydrocarbons and acids that help the quinones penetrate the outer waxy layer of the exoskeleton of predatory arthropods.

with claws equal in size, usually toothed or sometimes comblike. The abdomen has five segments visible below, the first three of which are fused together.

The tough, whitish, reddish brown to yellowish brown larvae are usually long and cylindrical or slightly flattened, with thick, tough body segments. The head is directed forward and is distinct from the thorax. The antennae are two or three segmented. Simple eyes are present, absent, or represented by dark spots. The five-segmented legs are well developed. The 10-segmented abdomen ends either with projections, spines, or dense hairs.

SIMILAR CALIFORNIA FAMILIES:

- ground beetles (Carabidae)—tarsi 5-5-5; first abdominal segment divided by hind coxae
- trout-stream beetles (Amphizoidae)—tarsi 5-5-5; found in or near water

- false darkling beetles (Melandryidae)—pronotum often with two impressions at base; first segment of hind tarsus longer than others
- pleasing fungus beetles (Erotylidae)—antennal club distinct; elytra sometimes with bright colored markings
- minute bark beetles (Cerylonidae)—antennae club distinct
- zopherid beetles (Zopheridae)—rough or ribbed elytra generally parallel sided in outline; antennae usually 10 segmented

Blister Beetles (Meloidae)

Blister beetles are of particular interest because of their medical, veterinary, and agricultural importance. They contain *cantharidin,* a chemical that causes irritation and blistering to sensitive tissues. In Europe, cantharidin was collected from the blister beetle called the Spanish Fly *(Lytta vesicatoria)* and taken orally for its purported qualities as an aphrodisiac. Cantharidin can be extremely toxic to humans, but as recently as the early 1900s, cantharidin was used to treat venereal disease, disorders of the bladder, bed wetting, and warts. Cantharidin is also poisonous to grazing animals that might accidentally consume blister beetles resting or feeding on plants. Some animals may learn to avoid large numbers of beetles brightly marked with contrasting colors of black, red, yellow, and orange. The feeding activities of some species of blister beetles can be ruinous to alfalfa and other field crops.

Cantharidin sometimes attracts antlike flower beetles (Anthicidae) of the genus *Notoxus.* These small beetles, with their antlike heads and distinctive horn projecting from their pronotum, are sometimes found swarming over dead blister beetles.

Adult blister beetles, if they feed at all, are strictly plant feeders, consuming leaves and flowers. They are often an important and conspicuous element of the desert fauna, where they are sometimes found swarming over the ground and their food plants by the hundreds or thousands. Blister beetles exhibit exceptionally diverse courtship behavior, using chemical, tactile, and visual cues. Eggs are typically laid on the ground or on plants. The larvae prey on grasshopper egg masses buried in the soil or consume subterranean nest provisions and brood of solitary bees. Blister beetles undergo a special type of complete metamorphosis known as hypermetamorphosis, with the larvae alternating between active and sedentary forms (see chapter 3).

CALIFORNIA FAUNA: Approximately 124 species in 17 genera.

IDENTIFICATION: Adult blister beetles (pl. 43; see also pl. 5) are black, metallic blue or green, or a combination of yellow, orange, red, and black, ranging in length from 3 to 35 mm. Their conspicuous, antlike head is directed downward and attached to the body by a distinct neck. The antennae are usually threadlike or beadlike, consisting of eight to 11 seg-

Plate 43. *Lytta magister* (Meloidae) (up to 35 mm) is a large, striking beetle found singly or in large aggregations in the Mojave and Colorado Deserts in the spring. Adults feed on various desert plants.

ments. The pronotum is without margins on the sides and narrower than both the head and the base of the elytra. The long, soft, leathery elytra are typically rolled over the sides of the abdomen. In some species the elytra do not meet in a straight line over the back or are slightly shorter than the abdomen. The tarsal formula is 5-5-4. The abdomen has six sternites visible from below.

The highly mobile first instar larvae, or *triungulins,* of hypermetamorphic blister beetles are tough bodied and silverfish shaped. The head is directed forward, with three-segmented antennae and zero to two pairs of simple eyes. The following four instars are C-shaped *feeding grubs,* with heads directed downward. The next stage, the *coarctate larva,* is virtually immobile and lacks legs and mouthparts. The seventh instar resembles the fifth grublike instar but does not feed and is followed by the pupal stage.

SIMILAR CALIFORNIA FAMILIES:

- soldier beetles (Cantharidae)—tarsi 5-5-5; elytra usually flattened, not rolled over abdomen
- false blister beetles (Oedemeridae)—head lacks neck

Longhorn Beetles (Cerambycidae)

Some of California's longhorn beetles, especially the males with their incredibly long antennae, are among the most spectacular and sought after of the state's beetles. They are particularly conspicuous in forested areas where they are attracted to freshly painted surfaces, cut wood, or lights at night. Longhorns have long been popular with collectors because of their bright colors, large size, and interesting habits.

Most California longhorns are nocturnal and are dull black or brown in color. They spend their days hiding beneath logs and bark, emerging at dusk or in the evening to search for food and mates. A few species are cryptically marked, allowing them to hide out in the open during the day, camouflaged against the bark of trees. By contrast, most day-active, or diurnal, species are flower visitors and are brightly colored. Some sport metallic blues or greens, whereas others are yellow or orange and seem to match the colors of their favorite flowers. A few species are bright red in color and, in the case of the milkweed borers *(Tetraopes),* this probably serves as a warning to birds and other predators that they taste bad (see chapter 3). Another eye-catching group of longhorns is the wasp and bee mimics, species boldly marked with black and yellow bands (e.g., the Lion Beetle *[Ulochaetes leoninus]*). The slender bodies and wasplike flight of this and other longhorns probably dupes potential predators into thinking that they can sting.

Adult longhorns feed on wood, leaves, and flowers. Wood-feeding species consume twigs, bark, bast (the fibrous inner bark), branches, trunks, and roots. Leaf feeders consume foliage, stems, needles, cones, fruits, and sap, whereas flower visitors devour pollen, stamens, and nectar.

The usually plump, cylindrical larvae feed upon the solid tissue of plants and are sometimes called roundheaded borers. They attack dead and decaying wood or living trees and shrubs, chewing their way into branches, trunks, stems, roots, and cones. Several species of longhorns girdle smaller twigs, whereas the feeding activities of other stem feeders may produce galls and other abnormal tissue growths. A few species prefer the stems of herbaceous plants. Root feeders usually tunnel inside, but some occasionally burrow in the soil and feed on the roots from the outside.

The excavation patterns of roundheaded borers are often used as diagnostic characters for their identification. The Ribbed Pine Borer *(Rhagium inquisitor)* tunnels and pupates between the bark and wood. Eucalyptus borers of the genus *Phoracantha* tunnel between the bark and wood but construct their pupal chambers in the heartwood. The Oregon Fir Sawyer *(Monochamus scutellatus oregonensis)* also tunnels between the bark and wood but prefers to feed and pupate primarily in the heartwood. Other species destroy only the heartwood and leave the outer, living sapwood intact. Trees hollowed and weakened by successive generations of borers, however, are easily toppled under their own weight or by windstorms.

The type of boring dust or frass that is pushed out of the tunnels of roundheaded borers during excavation is also characteristic of the resident species. As the larvae feed they produce powdery frass, flaky chips, or long, curly fibers resembling the excelsior used as packing material. Depending on the species, the tunnels are either tightly plugged with frass behind the larva or kept open. The newly emerged adult usually chews the exit holes. In a few species, the larvae make the hole and plug it with frass before pupating. Other insects, including leafcutter bees (Megachilidae), frequently occupy empty emergence holes.

Food preferences, or host plant specificity, varies among longhorn beetle larvae. Those with catholic tastes generally prefer wood that is either dead or decomposing but restrict their choice to either gymnosperms or angiosperms. Among those species preferring gymnosperms as hosts, many exhibit a decided preference for pines and firs or cypresses and junipers. The few species of longhorns developing in living wood are usually quite selective and are restricted to a particular species or genus of plant found within their range.

Female longhorns use their long ovipositors to place their eggs in or under bark or in wounds or cracks in the wood. Other species gnaw a slit in the wood to receive the eggs. Twig feeders place their eggs on leaf nodes, whereas root feeders choose the base of the plant near or below ground level. Because the eggs of all species require some degree of protection and moisture, few species deposit their eggs on exposed tree trunks without bark.

Some longhorns purposely create a suitable food source by girdling and killing stems. Adult females, sometimes with the assistance of the male, are responsible for carefully chewing a deep groove all the way around the stem. The dead, brown branch tips, known as *flags*, are especially conspicuous on evergreens such as coast live oak *(Quercus agrifolia)*. The girdle eventually weakens the branch, causing it to break and fall to the ground, where the larva continues to feed inside, undisturbed. In some species girdling is accomplished by the larva, which cuts off the branch's food and water supply by tunneling spirally inside the stem.

The natural enemies of California longhorns include other beetles, especially flat bark (Cucujidae) and checkered beetles (Cleridae). Immature eyed click beetles (Elateridae) routinely destroy the larvae of the Oregon Fir Sawyer and the Hairy Pine Borer *(Tragosoma desparius)*. Other predators include robber flies, assassin bugs, birds, and lizards. Woodpeckers are especially fond of roundheaded borers. Lizards congregate on fallen logs and branches to which the adults are attracted and capture them as they land. Spiders, scorpions, and toads also snap up nocturnal adults drawn to lights. Bats apparently attack large night-flying species such as the Pine Sawyer *(Ergates spiculatus)* and California Prionus *(Prionus californicus)*. Adults are parasitized by tachinid flies, whereas wasps, especially of the families Braconidae and Ichneumonidae, attack the larvae.

California's longhorn beetles play a beneficial role in the forest by recycling dead and dying trees, reducing them to humus. They can become serious pests, however, when managed timber supplies are weakened or killed by the ravages of storms, fires, or severe infestations of other insects. Much of the damage to timber is the result of an accumulation of roundheaded borers attacking the heartwood.

Stressed or injured shade and ornamental trees are particularly susceptible to attack by larval longhorns. With the exceptions of the California Prionus and the eucalyptus borers (*Phoracantha* spp.) (see chapter 5), the vast majority of the

Plate 44. The large and robust California Prionus (*Prionus californicus*, Cerambycidae) (24 to 55 mm) is widespread throughout western North America, except in the deserts. The root-feeding larvae reach 80 mm in length and are as thick as a man's finger. They feed primarily on the roots of living deciduous trees but also attack fruit trees, vines, and grasses. Adults are active in summer and are commonly attracted to lights.

state's longhorns are not considered important horticultural or fruit tree pests.

CALIFORNIA FAUNA: 309 species in 143 genera.

IDENTIFICATION: Adult California longhorn beetles (pl. 44; see also pls. 4, 11, 12, 17, 20) are extremely variable in shape. Most are elongate, cylindrical, or flattened, whereas others are more robust and broad shouldered. They range in length from a few millimeters to well over 60 mm. The head is directed forward or slightly to strongly downward and usually bears kidney-shaped eyes. Their antennae usually consist of 11 segments, occasionally more, and are usually inserted in the eye notches. The antennae are nearly always at least half the length of the body or more, especially in the males. The pronotum is almost square, oval, or cylindrical; the margin, if present, is spined or toothed. The scutellum is visible. The elytra are parallel sided and are smooth, densely punctured, or sparsely setose and almost always completely conceal the abdomen. Most species are winged, but a few species are flightless. The tarsi appear 4-4-4 but are actually 5-5-5. The fourth tarsomere is tucked away between the lobes of the heart-shaped third tarsomere. The claws are equal in size and usually simple. The abdomen has five or six visible sternites. Most longhorns are brown or black, but a few are metallic blue or green, mottled black, brown or gray, banded black and yellow, or with bright red or yellow elytra.

The white or yellowish white larvae, with dark heads, are typically elongate and cylindrical, but a few are distinctly flattened. The head of the larva is broad and retractable within the thorax and bears a pair of small, inconspicuous, three-segmented antennae and simple eyes. The thoracic segments are usually wider than the head and the abdomen. When present, the legs are very small and five segmented. The abdomen is nine segmented, with telescoping segments fitted with fleshy lobes to help them move through their tunnels. The last segment of the abdomen usually has one or two small projections.

SIMILAR CALIFORNIA FAMILIES:

- some click beetles (Elateridae)—mandibles enlarged; hind angles of pronotum extended backward
- soldier beetles (Cantharidae)—tarsi distinctly 5-5-5
- false longhorn beetles (Stenotrachelidae)—claws comb-like; tarsi 5-5-4
- leaf beetles (Chrysomelidae)—in some, the pronotum is always broadest in front; in others the antennae and body are short; in still others the elytra are broadest at the end

Leaf Beetles (Chrysomelidae)

Leaf beetles are one of the largest families of plant-feeding beetles in the world, second only to the weevils (Curculionidae). Some species are brightly colored or marked to indicate their distastefulness to predators. Both adults and larvae attack numerous kinds of plants, eating the bark, stems, leaves, flowers, and roots. Most species prefer flowering plants, but a few prefer conifers, ferns, and their allies. Most leaf beetles are specialists, feeding only on a single species of plant or groups of closely related plants. Even aquatic plants are not immune to their feeding ravages. The adults of some species eat leaves and other vegetative structures above water, whereas the larvae spend brief periods submerged underwater to feed.

Females of plant-feeding larvae lay their eggs directly on their host plants, either singly or in small groups. Some species apply a protective coating of their own feces on the eggs, sometimes using intricate rectal plates. Most larvae require living plant food to develop and spend much of their time grazing on the surfaces of leaves and other plant structures. Leaf-mining species tunnel between the upper and lower surfaces of living leaves, leaving discolored blotches, blisters, or trailing tunnels in their wake. In at least one California genus *(Monoxia)* the adults are also leaf miners. A few larvae feed on dried plant materials, such as dead leaves or bark on old twigs. Some are recorded to feed on ant eggs and waste materials and may even scavenge the bodies of dead ants.

Pupation generally occurs in the soil, sometimes in a cocoon, and usually in a cavity or special chamber dug by the larva. A few species may attach themselves to their host plants. The two California species of *Ophraella* pupate on their host plants inside a meshlike cocoon.

Leaf beetles employ various strategies to defend themselves against predators. Larvae grazing on the surfaces of leaves have evolved glands that shoot out of their bodies like party favors. The tissues of these glands are impregnated with noxious chemicals derived either directly from the host plant or produced by the larvae themselves. Tortoise beetle larvae and others live in cases constructed from their own fecal material or bits of plant detritus to look less appetizing to hungry predators.

Adult warty leaf beetles (*Exema* and *Neochlamisus*) also escape detection by resembling waste material. These small and short, chunky, dark beetles strongly resemble caterpillar feces and are no doubt overlooked by predators and collectors alike. Others may sequester the defensive chemicals of their host plants and incorporate them into their own defense system (see chapter 3). Some females cover their eggs with a protective coating of their own feces.

Many leaf beetle species are of economic importance. Garden and crop pests damage plants directly through their feeding activities or harm them indirectly by inoculating them with injurious diseases. Several pest species originating outside of North America have become established in California (see chapter 5), whereas others were intentionally introduced to control invasive weeds. The klamathweed, or St. John's wort *(Hypericum perforatum)*, is European in origin and infests hundreds of thousands of acres in the Pacific Northwest, including northern California. Displacing more desirable range plants, the toxic klamathweed causes mouth sores and intestinal distress in livestock. Based on research conducted in Australia, researchers at the University of California began testing two small, shiny leaf beetles, *Chrysolina hyperici* and *C. quadrigemina*, and a root borer, St. John's Wort Borer (*Agrilus hyperici*, Buprestidae), all of European origin. The first importation of the beetles in California arrived from Australia in 1944. Released in several localities in northern California, both species of *Chrysolina* were well established by 1948 and were quite successful in reducing or eliminating the pest weed throughout its range. *Chrysolina hyperici* has

Plate 45. Groups of Blue Milkweed Beetles (*Chrysochus colbaltinus,* Chrysomelidae) (6.5 to 11.5 mm) are often seen feeding and mating on the leaves of milkweeds, whereas the larvae probably feed on the roots. Disturbed beetles tuck their legs in and fall to the grasses below, producing a foul-smelling and foul-tasting fluid.

become widespread throughout the state, whereas *C. quadrigemina* appears to be restricted to the coastal mountains.

CALIFORNIA FAUNA: 475 species in 101 genera.

IDENTIFICATION: Leaf beetles (pl. 45; see also pl. 18) are extremely variable in shape. They are long and cylindrical to oval and convex or flattened, ranging in length up to 11 mm. Many species are uniformly black or brown, or iridescent blue, green, or bronze, or are two toned with distinctive patterns. The head is usually directed downward but is sometimes pointed forward. The antennae are usually 11 segmented and threadlike, saw toothed, or gradually clubbed. The pronotum is broader than the head and distinctly margined, sometimes broadly expanded and flattened. The scutellum is visible. The elytra completely conceal the abdomen; the margins are parallel sided or broadly rounded, sometimes broadly expanded and flattened. The elytral surface may have lines of pits or scattered hairs or appear shining and bare. The tarsal formula appears 4-4-4 but is actually 5-5-5, with the fourth segment small and hidden within the lobes of the third. The hind legs of flea beetles (e.g., *Altica* and *Disonycha*) are enlarged for jumping. Claws are usually equal in size and simple or with a single broad tooth. The abdomen has five segments visible underneath.

The larvae are also extremely variable and are humpbacked, caterpillar-like, spindle shaped, long and cylindrical, or somewhat C shaped. They vary in color from whitish to cream in subterranean forms to metallic or patterned in the exposed leaf feeders. The body, which may be covered with tiny spines, fleshy lobes, or just plain, may be covered or encased in fecal material or bits of plant detritus. The head is distinct and usually directed downward and has a pair of small antennae with one to three segments. One to six pairs of simple eyes are present or absent. The legs are three or four segmented. The abdomen appears to have eight segments visible from above, and the ninth and tenth segments are usually hidden. The abdomen may or may not end with projections.

SIMILAR CALIFORNIA FAMILIES:

- checkered beetles (Cleridae)—antennae gradually or distinctly clubbed; tarsi distinctly 5-5-5
- soft-winged flower beetles (Melyridae)—antennae saw toothed; tarsi distinctly 5-5-5
- some pleasing fungus beetles (Erotylidae)—antennae with distinct, flat clubs
- some handsome fungus beetles (Endomychidae)—tarsi appear 4-4-4, actually 3-3-3
- some darkling beetles (Tenebrionidae)—tarsi 5-5-4
- antlike flower beetles (Anthicidae)—head with neck
- some longhorn beetles (Cerambycidae)—pronotum usually not parallel sided, widest at middle or behind

Weevils or Snout Beetles, Ambrosia Beetles, and Bark Beetles (Curculionidae)

Curculionidae is the largest family of animals in the world, with over 60,000 species known to science. Most weevils are immediately recognizable by their mouthparts extending forward in a broad beak or outstretched into a long and curved snout. Either way, weevils are well equipped to chew their way through the leathery skins of seedpods or the tough shells of nuts. In contrast, the mouthparts of the wood-boring ambrosia and bark beetles are not extended at all. Weevils are generally hard bodied, and several species are wingless or nearly so and are incapable of flight. A few groups are capable of reproducing without a mate, a cloning process called *parthenogenesis.* In many parthenogenetic populations of weevils, males are entirely unknown.

Most Curculionidae feed on living plants and are associated with virtually every kind of aquatic or terrestrial plant and their various parts. Many larvae burrow into the stems of plants, whereas others feed on sick or healthy trees. A few are leaf miners or feed on dead leaves. The activities of a few root-feeding species may produce galls, or swellings on the roots.

Some adults attack fruits and nuts, but most feed on leaves, pollen, flowers, fungi, or burrow into wood. Adults may or may not feed on the same host as the larvae. The feeding preferences of many species makes them important pantry, garden, agricultural, and forest pests, including the Rice Weevil *(Sitophilus oryzae)* and Grain Weevil *(S. granarius)* and numerous bark beetles (e.g., *Dendroctonus, Ips*) (see chapter 5).

The majority of weevils are associated with flowering plants, but some are associated with conifers, especially pines.

The adults and larvae of generalist feeders consume a wide variety of plants, whereas specialists feed on a narrow range of suitable host plants. The larvae of generalists burrow through the soil and feed externally on roots, whereas the adults eat leaves. The larvae of specialists usually feed internally in the stems, roots, leaves, or the reproductive structures of one or more closely related species. The larvae of alfalfa weevils and their relatives *(Hypera)*, however, prefer to feed externally on leaves, flowers, or seedpods of their host. Specialists usually pupate inside the tissues of the host plant or in the soil near the base of the plant, but alfalfa weevils construct loosely woven cocoons attached to the host plant.

A great number of weevils associated with ornamental plants were introduced into North America from Europe in imported plants or amongst ballast brought by ships at the turn of the nineteenth century. Weevils infesting agricultural products are routinely intercepted at ports of entry into the United States and Canada. Others were deliberately introduced for biological control purposes, especially for the control of invasive weeds in western North American grasslands and southeastern wetlands. Exotic weevils have been introduced into California to control numerous plants, including scotch broom *(Cyticus scoparius)*, purple loosestrife *(Lythrum salicaria)*, gorse *(Ulex europaea)*, knapweed and starthistle *(Centaurea)*, and thistle *(Cirsium)*.

CALIFORNIA FAUNA: 570 species in 152 genera.

IDENTIFICATION: Adult Curculionidae (pl. 46; see also pl. 14) are broadly oval, long and cylindrical, to strongly convex beetles ranging in length up to 25 mm. Most species are black or brown, but a few species may be whitish, gray, green, or with metallic inflections. The head is directed downward, often with the mouthparts extended into a short, broad beak or stretched into a long and curved snout. The eyes are present, reduced, or absent. The antennae are usually elbowed

Plate 46. The Yucca Weevil (*Scyphophorus yuccae,* Curculionidae) (8 to 24 mm) is commonly found clambering among the spines and flower stalks of yuccas. Adults feed on the sap of living plants, whereas the larvae bore into the pithy centers of the flower stalks and crowns.

with a compact, three-segmented club. The pronotum is slightly wider than the head. The scutellum is small or absent. The elytra are rounded or parallel sided, bare or scaled, and almost or completely conceal the abdomen. The tarsal formula is usually 5-5-5, appearing or rarely 4-4-4, with claws equal in size and simple. The abdomen has five segments visible from below.

The plump and whitish, yellowish, greenish, or gray larvae are nearly cylindrical, slightly curved grubs usually covered with fine hairs and are without legs. The usually distinct head is directed downward and is seldom retracted inside the prothorax. The antennae are one or two segmented. The eyes are usually absent. The eight- to 10-segmented abdomen usually has three or four distinct folds or wrinkles on the back and lacks projections.

SIMILAR CALIFORNIA FAMILIES:

- pine-flower snout beetles (Nemonychidae)—antennae not elbowed; mouthparts with labrum; associated with male pollen-bearing flowers of pines
- fungus weevils (Anthribidae)—snout broad; antennae not elbowed
- straight-snouted weevils (Brentidae)—beak straight and pointed directly forward; antennae not elbowed or clubbed
- toothed snout weevils (Attelabidae)—antennae not elbowed; claws lobed

CHAPTER 7
STUDYING BEETLES

The study of beetles, either by observing and photographing them in the field, making a collection, or rearing them in the classroom, can provide a lifetime of exploration and pleasure. In a state of the size and natural diversity of California, new discoveries in the world of beetles are virtually guaranteed. Whether you live in the city or country, there are interesting beetles living in or near your home, wandering through the garden, or lurking about vacant lots and parks.

The basic study of beetles requires little in the way of material investment. All you really need to get started is a sharp pair of eyes, patience, persistence, and a bit of luck. As your fascination with beetles grows, so will your desire to know more about them. You can quickly expand your knowledge of beetles by learning where and how they live. Each kind of beetle has its own special requirements, habitat, and season. Your chances of locating new and interesting species are increased by searching a wide variety of habitats throughout the year, as well as employing various collecting techniques.

What Is the Best Season to Find California Beetles?

California generally has only two seasons: wet and dry. Much of the state has a Mediterranean climate, with a wet season that runs from September through April, with the heaviest precipitation occurring in December through February. The amount of rainfall generally decreases from north to south and increases with elevation. Good winter rains stimulate plant growth that triggers a flush of insect activity, especially of plant-feeding beetles, from late March through early June.

The sudden profusion of spring flowers in the Colorado and Mojave Deserts in March and April creates a haven for pollen and flower-feeding beetles, as well as those that scour the desert floor for dried bits of plants or animal remains.

Fresh piles of soil known as "push ups" mark the entrances of burrowing beetles as they prepare nests for their young or simply escape ever-increasing daytime temperatures. The steady succession of annuals and perennials flowering from April through June in the coastal sage communities of the Coast, Transverse, and Peninsular Ranges provides their beetle populations not only with food, but also with a place to find mates.

Spring is also the time when various rodents conduct their yearly nest-cleaning activities, especially in the foothills and inland valleys. As they remove plant refuse and waste from their underground chambers, ground squirrels and pack rats inadvertently eject beetles that make part or all of their living deep within the nest chambers. The beetles use their eviction to their advantage by taking to the air, sometimes by the hundreds, to search for mates and new burrows to occupy.

Aquatic beetles also begin to emerge from overwintering chambers beneath objects along streambeds, migrating to fresh pools in search of food and to reproduce. Shiny metal surfaces, such as those found on tin roofs, cars, and swimming pools, often attract these would-be aquatic colonists.

As plants of the deserts and foothills brace for the dry season (May to August), the mountains above 1,500 m (5,000 ft) in elevation begin to warm up. Mountain flowers attract different kinds of pollen- and flower-feeding beetles from those of the lowlands, whereas various species of wood-boring beetles take wing in search of mates and egg-laying sites. Isolated and intense summer thundershowers, typical of the montane regions of the Sierra Nevada and Transverse and Peninsular Ranges, spark the sudden emergence of still more beetles. The runoff from these storms feeds cold, fast-moving, mountain streams that are inhabited by aquatic beetles, whereas an entirely different set of aquatic species prefers the warmer, slower waters in the canyons and valleys below. In late summer (August and September), strong tropical storms occasionally push their way northward from Mexico to drench isolated patches of parched desert and the surrounding foothills, initiating a second flush of flower and beetle activity.

Fall can be a slow time of year for beetle collecting in California. The long summer drought that began five or six months earlier is nearing the end of its reign. To avoid the drying heat of late summer and the impending cooler temperatures of fall and winter, most California beetles have already sought refuge deep in the soil or beneath stones or logs. Naturally occurring stands of fall-blooming plants are uncommon in California. Expanses of rabbit brush *(Chrysothamnus)* sprinkled along the flats of the Mojave and Great Basin Deserts or dry desert streambeds lined with scale broom *(Lepidospartum squamatum)* attract insects of all kinds with their pungent yellow blossoms. These plants serve as the last gathering places for California beetles before the onset of the first frost. By late fall, most California beetles are decidedly less conspicuous.

During the winter months many adult beetles hibernate beneath stones or the loose bark of trees, sometimes congregating by the hundreds or thousands. Aquatic beetles may also gather en masse, seeking shelter in the moist mud or gravel beneath stones in otherwise dry streambeds.

On the other hand, a few California beetles are active in winter, springing into action only after the onset of cool winter rains. September marks the beginning of the wet season in northern California, whereas the southern part of the state must wait another month or more. While other parts of the country are in the grips of bone-chilling temperatures, much of coastal California remains frost free. Daytime temperatures in some inland valleys and the Colorado Desert can be quite mild and pleasant during winter, permitting further opportunities to observe and collect beetles. Coastal and desert dune-dwelling beetles move closer to the sand surface as temperatures drop and moisture levels increase. Even at higher elevations, with nighttime temperatures hovering around freezing, winter-active beetles may be found crawling slowly on the ground or across patches of snow well into the chilly night.

Where Are the Best Places to Look for Beetles?

Beetles are found everywhere in California, from rocky and sandy beaches upward to barren, windswept mountaintops thousands of meters above sea level. Although many beetles are habitat generalists with broad distributions throughout the state, others are specialists, inhabiting a single coastal dune, desert spring, or mountaintop. In terms of its climate and topography, California is more diverse than any other region of comparable size in North America. The variety of topographic, climatic, soil, and vegetation types that make up the landscape of California are woven together in a multitude of combinations to create a rich tapestry of potential niches, or microhabitats, available to beetles. Listed below are some of the more productive microhabitats to search for beetles.

Flowers and Vegetation

As a group, herbivorous beetles use all parts of plants not only for food, but also as places to mate and lay eggs. Look for them on all parts of shrubs and trees, especially flowers, fruit, cones, branches, leaves, and needles. A source of sweet nectar, high-protein pollen, and succulent petals, flowers are especially attractive to scarabs (Scarabaeidae), tumbling flower beetles (Mordellidae), metallic wood-boring beetles (Buprestidae), sap beetles (Nitidulidae), blister beetles (Meloidae), and longhorn beetles (Cerambycidae). Members of the sunflower family, such as brittle bush *(Encelia farinosa)* and coreopsis *(Coreopsis)*, are particularly attractive to flower-visiting desert beetles in spring. In summer, the blossoms of California lilac *(Ceanothus)*, buckwheat *(Eriogonum fasciculatum)*, lupine *(Lupinus)*, and numerous other flowers sprinkled over the mountains and their foothills often teem with beetles. Rabbit brush and scale broom act as magnets for fall flower visitors such as blister and longhorn beetles in the valleys and deserts.

Many flowers, shrubs, and trees are attractive to nocturnal species of beetles. Oaks *(Quercus)*, chamise *(Adenostoma fasciculatum)*, mountain mahogany *(Cercocarpus)*, buckthorn, mesquite *(Prosopis)*, buckwheat, pines *(Pinus)*, and firs *(Abies)* are all excellent plants to search after dark.

Dead Trees, Logs, and Stumps

Moist wood is usually more productive than completely dry wood. The quality of the wood changes as it decomposes and so does its desirability as a habitat. Checking the same stumps and logs over a period of years reveals a succession of beetle species that use the wood for mating, feeding, and egg-laying.

Night collecting on dead wood, particularly on warm evenings in spring and summer, can be very rewarding. Look for tumbling flower beetles, darkling beetles (Tenebrionidae), click beetles (Elateridae), longhorn beetles, and others emerging from their tunnels or wandering about in search of mates. During the day these beetles are found hiding underneath the bark. When pulling bark off of trees, be sure to examine the exposed area very carefully, checking both the log and the inner surface of the bark. Whenever possible, replace the bark when finished. Tie or nail the bark back to the log, and it will continue to attract and support beetles.

Freshly Cut and Burned Wood

The smell of freshly cut or recently burned wood attracts beetles, especially metallic wood-boring and longhorn beetles looking for mates and egg-laying sites. Boards or wood chips laid out across freshly cut stumps provide shelter for visiting beetles and should be checked frequently for new arrivals throughout the day and evening.

Fungi, Mushrooms, Mosses, and Lichens

Several families of small beetles may be found in good numbers on fungi, mosses, and lichens, and nowhere else. As with logs and stumps, inspect the plants and fungi carefully with a hand lens and leave them in good condition so that they continue to lure new beetles. If abundant you can take home samples of fungi, mosses, and lichens to extract beetles with a Berlese funnel (see below).

Stream Banks and Ocean and Lake Shores

Floating debris on the surface of streams and lakes contains flying and crawling beetles trapped by flowing waters. The flumes that criss-cross the western foothills of the Sierra Nevada regularly produce rare and unusual specimens. Ground beetles (Carabidae) and rove beetles (Staphylinidae) often hide during the day beneath plant debris that has been washed up on the shore of the ocean and lakes. Burrowing species, such as variegated mud-loving beetles (Heteroceridae), can be flushed from their burrows by splashing water across flat mud banks. Fast-moving adult tiger beetles hunt and fly along the shore, searching for insect prey.

Freshwater Pools, Streams, and Lakes

Although a few beetles prefer cold, fast streams, most favor ponds or slow-moving streams. Look for whirligig beetles on the surface of pools. Predaceous diving beetles are often found on gravelly bottoms or beneath submerged objects, while water scavengers and long-toed water beetles (Dryopidae) are found swimming near aquatic plants or crawling among mats of algae. Pick up and search rocks from flowing water for water pennies (Psephenidae) and riffle beetles (Elmidae).

Coastal and Desert Sand Dunes

Small scarabs, clown beetles (Histeridae), and weevils (Curculionidae) often hide in the sand among the roots of dune grasses and other plants, usually in the moisture layers or in accumulations of plant material beneath the surface. In fall and winter, the moisture layer and the beetles that live within it move closer to the surface and are easier to collect by hand or by sifting.

Carcasses

Dead animals provide food for many kinds of adult and larval beetles. In order to secure adequate food supplies for their young, burying beetles conceal smaller carcasses before other carrion-feeding insects arrive. Carrion beetles feed on fresh, juicy tissues, but skin beetles consume dried flesh. Hide beetles gnaw on hair, feathers, hooves, and horns. Predatory clown beetles, rove beetles, and ham beetles *(Necrobia)* search for the eggs of other carrion-feeding insects and mites. Other families of beetles are simply attracted by the moisture and shelter afforded to them by a dead body. Be sure to pick through or sift the soil directly beneath the body for all stages of beetles.

Dung

California's native dung beetle fauna is rather small when compared to other regions such as Arizona or states in the southeast. Large numbers of a few exotic species imported from Europe and Africa, however, may be found infesting cattle, horse, and dog dung throughout the state. Native dung beetles are often specialists, preferring the dung of burrowing rodents or deer. In the foothills of coastal southern California, hide beetles (*Trox gemmulatus*, Trogidae) are commonly found during the winter wet season gnawing on the hair-filled scats of coyotes. The shiny black Spotted Dung Beetle (*Sphaeridium scarabaeoides*, Hydrophilidae), a water scavenger beetle with red and tan spots introduced from Europe, is quite common in fresh cattle dung in the central and northern parts of the state.

Beneath Stones and Other Objects

Ground beetles, rove beetles, and darkling beetles may be found beneath stones, boards, and other objects lying on the ground. Stones in grass, along streams, or in other wet habitats are most productive. When finished searching the exposed area, always return the stone or board back to its original place for the benefit of the organisms living there, the aesthetic appearance of the area, and to maintain the productivity of the site for future collecting trips.

Leaf Litter and Compost

Layers of leaves and needles that gather beneath trees, accumulate along canyon bottoms, or wash up on beaches and lake shores harbor numerous beetles. Backyard compost heaps and other accumulations of decomposing vegetation are particularly productive. Some beetles, particularly some rove beetles, are found only under decomposing piles of seaweed along the beach.

Nests

Nests of birds and burrowing mammals may produce rare or poorly known species of beetles. Many small dung beetles prefer the waste of burrowing rodents. These beetles are active in spring and are sometimes found flying low over fields.

Be careful when searching these nests because there is a chance of contracting flea-borne diseases. Never disturb occupied nests, especially those of sensitive, threatened, or endangered bird or mammal species.

Lights

Beetles of many families are attracted to lights at night. Check porch lights, mercury vapor street lights, and store fronts, especially those in undeveloped areas. Although many species settle on the ground directly beneath the light, others are typically found climbing on nearby walls or plants. Others seldom come directly into the light, preferring instead to remain in the nearby shadows.

At Home

Beetles trapped in buildings usually fly to windows and other light sources while attempting to escape. Look for living and dead beetles on windowsills and light fixtures located inside houses, garages, sheds, and warehouses. High numbers of pest species may indicate an infestation of beetles attacking stored foods, skins, plant materials, or wood products.

Ethics of Beetle Collecting

Unlike most birds, butterflies, and dragonflies, which can usually be identified by sight alone, many beetles must be examined closely before an accurate identification is possible. Therefore, beetle collecting is essential to their study. Ill-informed bans on collecting divert attention from genuine conservation issues and ignore the benefits gained from responsible beetle collecting and associated scientific research. Responsible collectors are mindful of beetle populations and their habitats, never taking more specimens than they need.

Responsible collecting also helps ensure that land managers and the public view insect nets not as implements of carnage and destruction, but rather as tools of scientific advancement, environmental education, and ultimately, conservation. The very act of collecting beetles has sparked the careers of many scientists and teachers and provides important training that leads to a greater understanding of beetle classification, distribution, biology, and behavior.

The purpose of responsible beetle collecting is to develop and maintain a reference collection that can be used for both scientific and educational purposes. Beetle collecting by professionals and dedicated amateurs not only builds a record of beetle species living within a region, it also documents changes within beetle populations over time. This information provides an important historical record of species diversity in environments threatened by human activities or natural forces. Reference collections also provide the scientific basis for sound environmental policy. Amateur collectors and their collections, when working in concert with specialists and research institutions, are particularly useful in filling in gaps found in the collections of museums and universities and provide the basis for both scientific and popular publications.

When on a collecting trip, take only the specimens you need. Because of the great numbers of most beetles, you need not worry that your collecting activities will adversely affect most populations. Pesticide abuse, urban development, agriculture, automobile traffic, and bug zappers have a far greater impact on beetle populations then the actions of collectors ever could. Even competition and predation from introduced species take a greater toll.

The basic difference between vertebrate (bird, fish, reptile, amphibian, mammal) populations and those of beetles and other insects is their relative reproductive rates. Birds and mammals produce very few young in each litter or clutch and invest a considerable amount of effort into raising them. By contrast, each female beetle may produce hundreds of eggs, and the young seldom require parental care. Only a few survivors out of hundreds of eggs laid are required to successfully sustain most beetle populations. Beetles with small, sensitive, or specialized populations, however, such as those living in sand dunes, wetlands, vernal pools, or isolated plant populations or mountain ranges are particularly susceptible to changes in their environment. Repeated collections of beetles living in these habitats may, over time, adversely affect their population numbers.

The greatest threat to natural populations of organisms, including California beetles, is the modification or destruction of their habitat. Adopting a collecting ethic that not only recognizes the importance of maintaining beetle populations, but the preservation of their habitats as well, is essential. Field collecting should be selective and minimize trampling or other damage to the habitat and food plants. Collecting a large number of the same species at the same time or place seldom makes for a good reference collection. Rather, it is the diversity of a collection built over time, supported by accurate specimen labels and field notes, that creates a valuable scientific and educational tool.

Every time you go on a field trip, think of yourself as an ambassador for beetle collectors everywhere. Always ask for permission to collect on private and Native American lands and be respectful of other naturalists and their activities. When visiting public lands (such as county, state, and national parks, state and national forests, monuments, recreational areas, etc.) you must obtain written permission to collect beetles and other insects. Make sure all of your collecting activities are in compliance with regulations relating to the public and to individual species and their habitats. Managers of public lands are happy to issue permits to individuals conducting beetle surveys or other ecological studies, especially those affiliated with museums, universities, and other research institutions. Supporting beetle research is to their benefit because these efforts provide them with data they need to effectively manage and preserve habitats for all wildlife.

When importing and moving beetles, living or dead, comply with county, state, and federal agricultural and wildlife regulations. Generally speaking, moving any living beetle or other insect across county, state, or international borders requires written permission from state and federal agricultural authorities. Obtain all relevant permits in advance. The collection of beetles listed as endangered, threatened, or otherwise sensitive by state and federal wildlife authorities may be strictly regulated, and it is the responsibility of the collector to know and adhere to these regulations.

Equipment for Observing and Collecting California Beetles

Little in the way of specialized or expensive equipment is required to observe and collect beetles. To fully appreciate the sheer diversity of California beetles, however, some basic pieces of collecting equipment and methods will maximize your efforts in the field. Mastering these techniques and learning which ones work the best for a given beetle, habitat, or season is very worthwhile and will provide countless hours of satisfying activity.

Nets

Nets are useful for capturing beetles on the wing, resting on vegetation, or living in aquatic habitats. Aerial nets designed to catch flying beetles are usually lightweight, durable, and offer little air resistance when swung through the air. For tiger beetles that hunt on the ground but escape into the air when alarmed, quickly clap the aerial net over them. Then raise the tip of the bag and wait for the beetle to climb up into the net. Most beetle collectors prefer to use aerial nets with a rim diameter of 30 or 40 cm (12 or 15 in.) and a meter-long (3 ft) handle. The net bag should be reinforced along the rim with canvas or some other heavy material. The bag should be rounded, not pointed, at the bottom and be at least twice as long as the diameter of the rim so that it can be folded over the rim to prevent the escape of captured beetles. The net bag can be constructed of cotton bobbinet or some other soft translucent material that holds its shape. A commercially available net bag made of half canvas and half mesh is ideal as a heavy-duty aerial net or a light-duty sweep net.

Sweep nets generally have shorter, thicker handles, sturdy net rings, and net bags constructed completely of canvas to endure repeated brushing through dense vegetation. Sweep nets are carefully swept through the tops of grasses, shrubs, or tree branches. By keeping the rim of your net vertical to the ground, you increase the chances of knocking beetles perched on plants into the net bag. After making several sweeps in this manner, swing the net back and forth a few times to make sure all of your captures are driven to the end of the bag. Then flip the bag over the rim of the net to prevent their escape. Open the net bag carefully to release stinging bees and wasps, then remove beetles from the net either by hand, with forceps or an aspirator (see below), or by capturing them inside the net with a killing jar.

Long-handled dip nets or aquarium nets are useful for capturing beetles in ponds and streams. Swimming beetles may simply be scooped up out of the water. D-frame nets have a rim with a flat side. They can be dragged along the bottom to dislodge and capture beetles resting on rocks and plants. D-frame and dip nets placed vertically against the substrate readily capture beetles dislodged by lifting rocks and disturbing vegetation immediately upstream.

Beating Sheet

Beating is one of the most productive methods for collecting beetles from the foliage and branches of trees and shrubs. Beating sheets are usually 1 m^2 (9 ft^2), with reinforced pockets sewn into each corner. The sheet is supported by two hardwood dowels or plastic tubes, each spanning the opposing corners in the form of a cross.

Place the beating sheet beneath the foliage, then strike the larger branches over the sheet with another dowel or a net handle. Beetles and other insects lose their footing and fall onto the sheet where they can be picked up with a pair of forceps or an aspirator. During warm spring and hot summer days when beetles are likely to escape quickly, beating is most effective during the cooler morning or evening hours when they are unable to move as fast.

Extracting Beetles from Soil and Plant Debris

Kitchen strainers with screened bottoms of various diameters are useful for sifting through sand and fine, dry soils at the beach, dry river bottoms, or desert sand dunes, especially at the bases of plants. The screen mesh should be large enough to let most of the sand flow through, but not the beetles and larger bits of debris. Pick the beetles from the debris with forceps or extract them by means of a Berlese funnel.

The Berlese funnel is a device that uses light and heat to drive beetles from a sample of compost, beach wrack, leaf litter, rotten wood, or other plant debris into a collecting container. Place a large funnel that narrows to about a 1 cm opening, small side down, into a tall glass jar with a wad of moist paper towel in the bottom. Fill the funnel with leaf litter, and place a high-wattage light bulb directly over it. As the litter warms and dries, beetles and other arthropods move deeper into the detritus as they seek cooler, wetter conditions. Eventually they fall down the funnel and into the collecting jar. Depending on the size and moisture content of the debris, it may take several hours or even days to drive all of the beetles from the sample.

Another useful method for separating beetles from plant materials—particularly clumps of grass, but also fungi, bark, and dung—is to drop them into a bucket of water. Beetles and other insects float to the surface where they can be scooped up with a small kitchen strainer or collected by hand.

Night Collecting

Searching for beetles on warm (65 degrees F or higher) spring or summer nights, especially nights with little or no moon, can be very fruitful and can produce species different from those collected during the day. A headlamp allows you to illuminate your nocturnal searches while keeping both hands free to collect beetles crawling on the ground, hiding beneath objects on the ground, or wandering about plants and logs.

The most productive method of collecting beetles at night involves a light suspended in front of or over a white sheet. Although almost any light attracts night-flying beetles, ultraviolet lights, or black lights (pl. 47) are the most effective. Suspend a freshly laundered bed sheet between two trees or poles over a ground sheet. Then place a black light about one-third meter from and parallel to the upright sheet at about eye level to achieve maximum illumination. Not all beetles come directly to the sheets, so be sure to regularly patrol the perimeter of the lighted area, both on the ground and on nearby plants, for longhorn beetles and other species that may prefer to remain in the shadows. Run the light for many hours to take advantage of the succession of different species drawn to the light throughout the night.

Commercially available black lights operate on house current or 12-volt batteries. Mercury vapor lights using 175 watt bulbs are also attractive to beetles and other night-flying

Plate 47. The most productive method of collecting beetles at night involves an ultraviolet or black light suspended in front of or over a white sheet. Place a black light one-third meter from and parallel to the upright sheet at about eye level to achieve maximum illumination. Photo by C. L. Hogue.

insects. A word of caution, though. Mercury vapor bulbs become extremely hot and break should they come into contact with rain or other sources of moisture. Mercury vapor lights require house current or a generator to operate.

Trapping Beetles

Several methods can be used for capturing beetles without being present in the immediate vicinity. Pitfall trapping is useful for collecting beetles that crawl on the ground. Using a trowel or small shovel, dig a small hole just large enough to accommodate a 16-ounce plastic deli cup. Place the cup in the hole so the rim is flush with the surface of the ground. Then place another deli cup of the same diameter inside the first. By nesting the containers in this fashion you can easily check your traps without having to redig the hole each time you reset them. Cover the trap with a flat stone or slab of wood raised on small stones, leaving a space large enough for beetles to crawl into the cup but small enough to keep out larger animals.

Place traps in a variety of habitats to increase the diversity of your catch and check them daily. Pitfall traps can be baited with small amounts of fresh dung, carrion (use beef, chicken, or fish), rotting fruit, or chopped mushrooms. Wrap the bait in cheesecloth and suspend it over the opening of the cup. Mixtures of equal parts of molasses and water, or malt and yeast, attract sap-feeding beetles. Try different materials to attract different species of beetles. Keep in mind that much of California is dry and that some baits must be changed frequently to maintain their moisture levels and drawing power.

Bait is not always necessary. Unbaited pitfall traps can be placed along natural barriers such as rock ledges. Strips of wood, metal, or plastic act as runners leading to the trap and increase its effectiveness. A small wad of paper towel or crumpled leaves in the bottom of the container gives trapped beetles a place to hide from one another. Check your traps frequently.

For traps left out for a week or more at a time between inspections, pour a few centimeters of propylene glycol antifreeze into the bottom of the pitfall trap. Propylene glycol preserves beetles that have fallen into the trap and is not toxic to mammals.

Another method for attracting ground-dwelling scavengers, especially at night in deserts or along paths or other open habitats, is to leave a trail of oatmeal. Check your oatmeal trails regularly because there is nothing to keep the beetles around except for their appetites.

Pan traps can be made out of any shallow, bright yellow plastic container and filled with several centimeters of water. The bright yellow color attracts flower-visiting beetles and other insects. A drop or two of dish soap added to the water breaks the surface tension and makes it harder for the beetles to escape.

Light traps use a black light suspended over a funnel placed on a five-gallon bucket. By supplying the bucket with plenty of wadded-up paper towels you can reduce the wear and tear on beetles and other insects caught in the trap.

Flight intercept traps (pl. 48) are extremely useful for sampling beetles in a given habitat and frequently capture species seldom encountered otherwise. These traps are especially effective for collecting small, crepuscular species. The trap consists of a dark mesh nylon screen suspended between two poles and placed across a trail, next to a log, or in a field. A shallow trough or a series of roasting pans containing soapy water or propylene glycol is placed directly below the screen. Most beetles flying into the screen fall down into the fluid where they are killed and temporarily preserved.

Tools for Handling Beetles

There are several ways to handle and manipulate living and dead beetles easily without damaging them. Forceps made of spring aluminum known as "featherweights" are extremely useful for picking up small beetles, whereas camel hair

Plate 48. Flight intercept traps are extremely useful for sampling beetles in a given habitat and frequently capture species seldom encountered otherwise. These traps are especially effective for collecting small, crepuscular species. Most beetles flying into the screen will fall down into the roasting pans filled with soapy water or propylene glycol.

brushes are used to probe for and dislodge beetles from their resting and hiding places.

Aspirators are simple tools that suck small beetles inside a glass or plastic vial. There are several designs, but the most widely used model incorporates a close-fitting rubber stopper that fits into a transparent glass or plastic vial. Two copper tubes pass through the stopper. The collecting tube is open at both ends and is used to collect specimens. The outer portion of the second tube is fitted with a flexible section of surgical tubing that serves as mouthpiece, whereas the inner section protruding into the vial is covered with fine mesh gauze. Small beetles are sucked through the collection tube and into the vial by drawing air through the mouthpiece. Although the protective gauze prevents the accidental inhalation of a beetle, it does not protect you from molds, spores, and the noxious odors and chemicals produced by some beetles. Models with a suction bulb instead of a mouthpiece alleviate these potential hazards.

Killing Jars and Agents

Beetles retained for the purpose of making a collection are transferred to a killing jar where they are dispatched as quickly and humanely as possible. Any jar can serve as a killing jar as long as it has a tight fitting, screw-top lid. A killing jar can be any size but should slip easily into a pocket or collecting bag. Beetle collectors generally use long, cylindrical vials, such as those used for olives, pickled onions, or spices. These jars range in length from approximately 7 to 15 cm (3 to 6 in.). Even the largest beetle in California, the Pine Sawyer *(Ergates spiculatus)* (pl. 4), easily fits in a jar 4 cm (1.5 in.) in diameter. The screw-cap can be replaced with a tight-fitting neoprene stopper for ease of removal and retention of the killing agent. Be sure to fill your jar with loosely wadded paper toweling to hold the killing agent and to absorb any fluids produced by your catch.

Ethyl acetate is the safest chemical to use and is available from biological and scientific supply houses and some pharmacies. Although it is relatively safe to use, avoid getting ethyl acetate on your skin or breathing the fumes. Add several drops of ethyl acetate until the paper towel is moist, but not wet. Continual opening and closing of the jar results in the loss of fumes as the ethyl acetate evaporates, so the killing jar will have to be recharged from time to time. Keep in mind that ethyl acetate dissolves hard plastic bottles (e.g., most pill bottles) made of styrene, as well as styrofoam.

Some collectors prefer to use killing jars with plaster of Paris on the bottom to absorb the ethyl acetate. First select a number of jars whose mouths are large enough that you can reach in with your fingers to clean them from time to time.

Stuff a 10 mm thick (.5 in.) layer of cotton or sawdust onto the bottom of each jar. Then pour a smooth, 5 to 10 mm thick (about .25 to .5 in.) layer of plaster of Paris in each jar. Let the plaster of Paris dry and harden overnight. Prepare jars of different sizes to determine which works best for you.

Before using the jars, pour a teaspoon or two of ethyl acetate into the jar and let it stand long enough for the liquid to become absorbed, then pour off any excess. Keep the jar closed as much as possible because the fumes are extremely volatile and evaporate quickly when the jar is open. Add loosely crumpled tissue to the jar to absorb excess moisture from the ethyl acetate and your specimens. Never put beetles in a killing jar with other orders of insects because they will damage other softer-bodied insects with their mandibles and claws before dying. Butterflies and moths will cover beetles with scales, making them unattractive and difficult to identify.

Temporary Storage of Specimens

Under ideal conditions, beetle specimens should be mounted and spread immediately. Several methods, however, can be used for storing specimens temporarily. For short periods, beetle specimens can be left in the killing jar until you are ready to mount and spread them. Assuming adequate levels of ethyl acetate are maintained in the jar, your specimens will remain pliable for a few days and can be handled easily without damaging them.

Specimens may be transferred to another container and sealed to keep in moisture and stored temporarily in the refrigerator or freezer to maintain flexible legs and antennae. Specimens kept in containers that are not airtight soon become dry and brittle in a frost-free refrigerator.

For longer storage, translucent, 35 mm film canisters with tight-fitting lids are readily available and easy to store. Simply place the beetles into the container while they are still pliable, leaving enough room at the top for a folded square of toilet paper moistened with a few drops of ethyl acetate. Specimens stored this way keep indefinitely, but delicate colors may fade and tufts of hairs become matted.

Another method for temporarily storing specimens is to place them between layers of Cellucotton or in paper or glassine envelopes. They may then be stored in a small, tightly closed box, preferably with napthalene moth flakes. The Cellucotton prevents damage and distortion from pressure and keeps the beetles from tumbling around during transport. One disadvantage of this method is that specimens dry out before they are mounted. Drying out can be avoided by adding chlorocresol crystals to the storage box. The crystals keep the specimens relaxed and eliminate the need for a relaxing chamber. Chlorocresol can be purchased from a biological supply house.

Vials of 70 percent alcohol with neoprene stoppers are used to store large numbers of beetles collected in Berlese funnels, pitfalls, and black light traps. Beetles kept in alcohol remain relaxed, but specimens with delicate colors stain or fade, whereas tufts of hair become matted. Be sure to include basic collecting information (locality, date, collector) with each container, using pencil or permanent ink on good-quality, acid-free paper.

Viewing Specimens in the Field

No one who spends any time in the field should be without a good quality hand lens. Available from biological supply companies, hand lenses are small and compact devices for revealing beetle anatomy and other details that might otherwise escape notice by the naked eye. Magnifications of 8× or 10× are ideal, with some units employing several lenses used in concert to increase magnification. The trick is to hold the hand lens close to your eye and then move in on your subject until it comes into sharp focus.

Records and Field Notes

Always record the date, place, and collector for your specimens. These data become the basis for labels that accompany each specimen in your collection. For locality records in California, always note the county and the distance from important road junctions or towns. Next, determine and record the latitude and longitude of the locality. With the availability of inexpensive yet accurate global positioning systems, obtaining this information is as easy as pushing a button. The California atlases and gazetteers, published by the DeLorme Mapping Company (see "Selected References"), provide excellent topographic maps and are easy to use for obtaining locality coordinates. When recording locality notes, especially for use on specimen labels, always consider them as directions. Put yourself in someone else's shoes. If you needed directions to find a locality to search for a particular beetle, what information would you require?

Remember that dead specimens in collections reveal little of the beetles' lives. These important details are only discovered through careful observation of living beetles. Therefore your field notes should include time of day, temperature and humidity, plant or animal associations, behavior, or other important facts. Anything that catches your eye is worth recording and may easily prove to be new to science. On what plant was it feeding or laying its eggs? What postures did the beetles assume during the act of mating? How many individuals did you see? Were males more abundant than females? Did you observe other animals preying on the beetles? Did you find beetle remains in animal scats? Everything is fair game when it comes to making notes. But be selective, otherwise you will spend all your time writing instead of seeing! With some experience, you will settle on a standard of how and what you write. Whenever possible, record your observations in the field as they are happening. Never trust your memory for long because it is all too easy to confuse bits of information in time and place.

Remember that the value and quality of your records and observations are greatly enhanced by a clear association with identified voucher specimens in a collection. The best notes are those fully integrated with a well-curated collection. In the absence of reliably identified specimens, careful descriptions, including drawings and photos, may be sufficient. Careful note taking and rigorous record keeping not only supply our basic knowledge of beetles, it also allows you and others to go back and repeat your observations.

Choose a notebook or journal of acid-free paper that is sturdily bound for rugged use in the field. It should be easy to pack, but not so small that it is easily lost or misplaced. And it should lay flat for ease of use. Loose-leaf notebooks work well because they can be left safely at home, and pages filled out in the field can be added as they are generated. Although a number of writing implements are available on the market today, fine-tip marking pens with permanent ink or pencils are the most reliable for writing under a variety of conditions.

Making a Beetle Collection

Responsible beetle collecting requires that you take no more specimens than you need and certainly no more than you are able to properly mount, label, and safely store. Because well-curated collections are important scientific and educational tools, it is essential that all specimens are protected from the physical damage and deterioration caused by light, molds, and insect pests.

Collections should include permanently recorded field notes detailing habitat, weather conditions, and other pertinent information. In fact, keeping notes of your observations of behavior and other biological interactions should be considered equally important to collecting. Photographic records or drawings, with full data, are also encouraged. Whenever possible, use all the elements of your collection to educate the public as to the reciprocally beneficial activities of collecting and conservation. Also, volunteer your data and donate extra specimens to museums, universities, and other researchers who are actively publishing their results in scientific journals. These contacts dramatically increase your enjoyment of beetles and contribute to our overall understanding of them.

Although making a beetle collection is a highly personal endeavor, it is important to consider that your collection, if properly cared for, will last hundreds of years, inspiring and informing future generations of researchers and naturalists. Think of yourself as a temporary caretaker of your collection; ultimately it belongs to science. In the event that you lose interest, lack adequate storage space, or simply want to know that your hard work will be preserved long after you are gone, consider donating your collection and its associated records to an appropriate research institution. These institutions are in the business of caring for collections and will make your contribution to science available to the entire community in perpetuity.

Mounting and Spreading Beetles

The procedures used by insect museums and collectors in North America for mounting and labeling beetles are designed to encourage neatness, conserve space, and assure easy examination of all necessary characters without damaging the specimen.

Sewing pins are too short and corrode easily. Always use insect mounting pins, available through suppliers, for your specimens. Insect pins are longer than sewing pins and are available in a variety of sizes to accommodate specimens of different sizes. They are also coated in varnish to prevent rust. Be sure to use insect pins with secure nylon heads because the pins made with brass heads tend to slip down the shaft of the pin. Sizes 0, 1, 2, and 3 work for most California beetles.

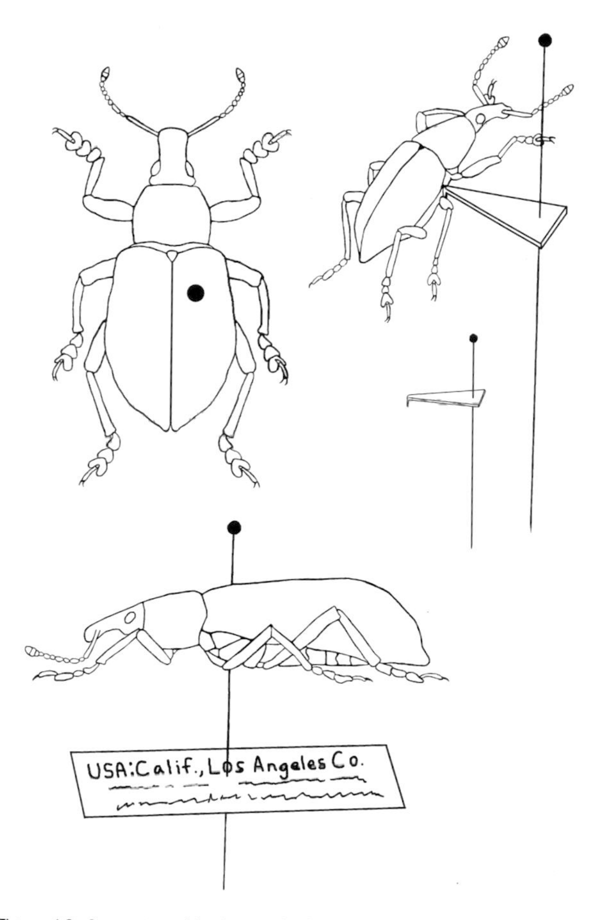

Figure 16. Correct positioning and orientation for pinning or point-mounting beetle specimens.

Beetles are always pinned through the right elytron so that the point comes out beneath the body between the second and third leg (fig. 16). Specimens can either be held firmly by thumb and forefinger or placed on a smooth, soft surface that will accept a pin, such as a styrofoam block covered with paper. Mastering this skill takes practice and it is not unusual for beginners to drive the pin through the specimen's leg or tear it off altogether. Before driving the pin all the way through, examine the specimen carefully from the side and head-on to determine that the final position of the pin is at right angles to the beetle's line of center from these points of view. Push the pin through your specimen, leaving 8 to 10 mm (about .4 in.) between the top of the specimen and the head of the pin.

The legs and antennae of the beetle must be positioned so all necessary characters needed for identification are easily visible. To identify your specimens you may need to count the number of antennal, tarsal, and abdominal segments, inspect leg bases and ventral abdominal sutures, and examine the features of the eyes and mouth. Once you become familiar with the characters you need for identification, mounting your beetle specimens properly will become second nature. For display purposes the wings of beetles can be spread like those of a butterfly with the aid of a spreading board. Although this may look interesting, the time and effort expended here has little scientific or practical value.

Beetles that are too narrow to pin on a no. 0 insect pin are placed on points (fig. 16). Points are triangles made from acid-free card stock and are about 7 mm long and 2 mm wide at the base. Points are cut with scissors or made by a point punch, a commercially available device similar to those used for checking tickets. A no. 2 or 3 insect pin is pushed through the broad end of the point. The narrow end of the point is dipped in a small droplet of glue. Clear nail polish works well for this purpose. The beetle specimen is then carefully glued to the point on its right side near the second leg. Sometimes the tip of the point is bent down at a right angle with a sharp pair of forceps so that once the beetle is attached, its dorsal surface faces upward rather than toward the pin. Align the long axis of the point so that it is centered on and parallel to the locality label.

Mounting Blocks

These simple devices are used to achieve equal spacing on the pin between the specimens and head of the pin and its labels. Mounting blocks may be purchased or made with a block drilled with four fine holes. Each hole should be drilled in 6 mm (.25 in.) increments, beginning with 6 mm.

Labeling

Now that personal computers are common at home and in the office, neatly printed specimen labels are easy to produce. Labels should only be produced with laser printers on acid-free card stock. Acid-free card stock and papers will not become yellow and turn brittle, protecting your label data from the ravages of time. A font size of four or five points will keep the labels small, yet legible. Ideally, a basic locality label should have no more than five lines and identify the country, state, county, general locality, specific locality, latitude and longitude, elevation, date, and collector. To avoid confusion completely, spell or abbreviate the actual name of the month. A reference number to photographs or field notes can be incorporated in this label or can be placed on a separate label. A sample locality label is given below:

USA: CA, Santa Barbara Co.
Skofield Park, Rattlesnake Cyn.
Santa Barbara, 34.45°N 110.23°W
elev. 120 m, 12 August 2004
A.V. Evans & J.N. Hogue

Additional labels may be used to include method of collection, host plant, or other ecological data. Once the beetle is identified, a label containing its name may also be used. Each label should be pinned below the specimen lengthwise so that it is centered on the specimen itself, regardless of its position on the pin. The left-hand margin of the label should be oriented toward the head of the specimen.

Relaxing Specimens for Mounting or Dissection

Dry, brittle specimens must first be relaxed before they can be handled without damaging delicate appendages. A number of methods can be used to reconstitute the moisture levels of specimens, rendering them more pliable. One way is to simply drop them directly into boiling water with a bit of dish soap added as a wetting agent. After a few minutes the specimen should become pliable. Large, heavy-bodied beetles may take longer. Another method is to create a sealed chamber with a high level of humidity inside. Using a plastic shoebox as the chamber, place a layer of clean sand, cardboard, or some other relatively sterile, porous material at the bottom of a box that holds moisture. Soak the material with water and pour off the excess. Place dry specimens on a plastic lid so they are not touching any surfaces that may become wet. Add a moth ball or two to the chamber as a fungicide. Remember that hard, clear plastics such as styrene and styrofoam dissolve when coming into contact with moth balls, a reaction that could damage your specimens. Although most beetles become sufficiently relaxed in the chamber overnight, it may take several days to adequately soften larger specimens for mounting or dissection.

Preserving Larvae and Pupae

Specimens of larvae and pupae, especially those with ecological data and positively associated with the adults, are extremely valuable. Immature beetles are killed in 70 percent ethanol or isopropyl and then stored in glass vials with screw caps or neoprene stoppers with a fresh supply of alcohol. Specimens killed in this way, however, may become discolored and distorted, making their study difficult. Better results are obtained by first placing the immature beetle in boiling water to kill the organisms and fix its tissues. Then place the specimen in a vial of alcohol. As with pinned specimens, make sure that each vial is properly labeled.

Storing, Arranging, and Caring for Your Beetle Collection

Sturdy, airtight collection boxes kept in cool, dry rooms are essential for maintaining and managing a beetle collection. Over time, sunlight fades specimens, and warm moist air fosters mold that destroys specimens. Museum pests, such as skin beetles and book lice (Psocidae), scavenge dead insects for a living. These small insects can slip into the smallest of openings and will reduce beetle collections to nothing more than pins, labels, and dust. Tightly sealed boxes or boxes regularly supplied with moth balls or crystals (paradichlorobenzene) keep out most pests but may not kill them once they have become established. Placing entire collections (boxes and specimens) in a freezer for up to a week or more will kill immature and adult museum pests.

Any tight-fitting box will serve to house a collection as long as it is at least 5 cm (2 in.) deep with the lid closed to prevent squashing or knocking over pinned specimens. A soft pinning bottom 6 to 10 mm (.2 to .4 in.) thick will hold the pins in place. A layer of cork or fiberboard is suitable if it is not too hard, but a sheet of polyethylene foam works best. The stryofoam used to secure items in boxes or sold in florist and hobby shops works well, too, but will melt if it comes into contact with moth crystals used as an insect repellent. Cardboard also works as a pinning bottom, but it is sometimes difficult to use with the smaller sizes of insect pins. A number of commercially available insect storage boxes and drawers to suit all budgets are on the market. Occasionally you can buy used specimen boxes at reasonable prices from the entomology departments of museums or universities.

As your collection grows you might want to consider a more permanent and compartmentalized approach. Most entomology departments use a system of tightly sealed glass-topped wooden drawers stored in sealed metal cabinets. Instead of pinning bottoms, each drawer contains a series of interchangeable paper boxes of various sizes with polyethylene foam bottoms. These boxes, or unit trays, allow you to easily organize your collection and add new specimens.

Pinned specimens are aligned into neat columns and rows, either by label or specimen, whichever is the most conspicuous. The head of each specimen should face away from the observer. Never crowd specimens so they cannot be easily removed from the box without entangling the antennae and legs of their neighbors. Arrange pointed specimens in a similar fashion so that the narrow end of the point with the specimen is facing away from the observer, with the head pointed to the right.

Collections of immature and adult beetles kept in alcohol should be kept in sturdy boxes where they are easily accessible and away from extreme temperatures. Display collections must be kept away from extreme temperatures and out of direct sunlight. Display cases fitted with UV-filtered plexiglass slow but do not prevent the fading of specimens due to sunlight.

Examination and Identification of Beetle Specimens

Most beetles are small, so their identifying characteristics must be viewed with magnification. A hand lens is useful, but a stereoscopic dissecting scope is ideal. Be sure to view the specimen from various angles before deciding the state of a particular feature, especially when determining the number of foot or antennal segments.

Many beetles can only be positively identified to species through examination of the male reproductive organs. With a bit of practice you can learn to extract the genitalia from the posterior opening of the abdomen by gently pulling them out with fine-tipped forceps or with the aid of a fish-hooked insect pin. This is best accomplished with freshly killed specimens. Dried specimens must be left in a relaxing chamber or placed for a few minutes in boiling water with a drop or two of dish soap added as a wetting agent. The genitalia are either left attached by their own tissues just outside the abdomen to dry in place or removed entirely and glued to a point and pinned immediately below the specimen for examination.

Keeping Beetles in Captivity

Working with live beetles in captivity provides an important dimension to your studies and allows you to observe behaviors for which pinned specimens remain forever mute. Besides, they make excellent educational displays and are a surefire way of bringing life to a discovery room, classroom, or nature center. Captive beetles also provide a means of acquiring pictures of species or behaviors that are difficult or impossible to photograph in the field.

Bringing 'Em Back Alive

The first step is to collect and transport beetles so that they arrive at their final destination alive and unharmed. Captive insects exposed to sudden increases in temperature die quickly, especially aquatic species. Never leave captive beetles in direct sunlight or unprotected in a closed car for any period of time.

Carry an ice chest with one or more two-liter bottles filled with frozen water for transporting live beetles from the field. Half-pint and pint-size plastic deli containers, available at supermarkets and restaurant supply chains, are ideal containers for housing beetles because they are inexpensive, lightweight, and stack easily for packing. Punching air holes in the lid for a day trip, especially if the containers are kept cool, is not necessary. A piece of paper towel, some leaf litter, or a piece of moss, slightly moistened with water, provides a bit of comfort for your animals and protects them from the jostling of travel. Never transport aquatic beetles in water; they will drown. Instead, use wet moss or moistened paper towels as a substrate.

Rearing from Wood

Late winter and early spring are good times to gather fungi, dead limbs, and rotten logs and stumps that contain beetle larvae and pupae. A large mouth glass jar can serve as a practical and inexpensive rearing chamber. Supply your rearing material with a bit of water from time to time to keep moisture levels up. Use cheesecloth secured with a rubber band or window screen glued to a dome lid ring of a canning jar as a top to allow excess moisture to escape. Be sure to place your jar away from outside doors, windows, and heating and cooling ducts to avoid exposing your animals to extreme temperatures. The warmer indoor temperatures accelerate their development.

For larger rearing operations, place dead limbs in sturdy plastic sweater boxes or large, square metal cans. Seal the containers tightly to prevent the escape of beetles when they emerge from the wood. Wrap the plastic boxes with black garbage bags to keep out light. Then cut a hole on the end of each container and insert a small jar or vial. Emerging beetles are drawn to the light and find their way into the bottle. Be sure to add a bit of water from time to time so the rearing material does not completely dry out.

Housing for Adult Beetles

To keep beetles in captivity it is important to know what they need in terms of food and water. With these basic requirements in mind, you have a better chance of duplicating their environment as much as possible and ensuring that they thrive and behave as naturally as possible. Based on your observations in the field, add rocks or bark for shelter, twigs or branches for climbing, and soil or leaf litter for egg laying. Almost any plastic or glass container found around the house can be transformed into a home for terrestrial beetles. Make sure that the container has good ventilation. Beetles and other insects do not need much oxygen, but fresh air is required to release heat and control humidity. The presence of mold is a clear indication of poor ventilation combined with too much moisture and may be harmful to your animals. Offer water regularly by misting the container with filtered or distilled water or by placing a vial filled with water and plugged with a wad of cotton in the cage.

Many plant-feeding species are specialists and require fresh cuttings of their favorite foods. Beetles with more generalized tastes may be offered a variety of plant foods: leaves, lettuce, oatmeal, potato slices, and various kinds of fruit. Remove uneaten food after a few days to prevent the buildup of mold, mites, and gnats.

Many California beetles are easily kept in captivity. Adult and larval mealworms *(Tenebrio molitor)* are available in pet stores and bait shops. Darkling beetles *(Eleodes)* lead long, active lives in a terrarium supplied with several centimeters of soil, dry leaf litter, and the occasional pinch of flaked fish food or oatmeal.

A Green Fig Beetle *(Cotinis mutabilis)* (pl. 49) captured during the late summer can be kept in a terrarium supplied with 5 to 8 cm (2 to 3 in.) of potting soil and branches to climb. It feeds on a variety of soft fruits including peaches, figs, bananas, and grapes. It reproduces readily, and its C-shaped grubs soon appear in about a month or so in the soil. It thrives on grass cuttings, leaf litter, and ground-up dog food. The larvae are usually present year-round in well-established compost heaps.

Predatory beetles, such as ground beetles, are long-lived in captivity, although some species are somewhat secretive in their activities, usually becoming active only at night. Cater-

Plate 49. The Green Fig Beetle, or Peach Beetle (*Cotinis mutabilis,* Scrabaeidae) (20 to 30 mm), is common in the coastal plains of southern California, portions of the Antelope Valley, and the length of the Central Valley. Adults feed on a variety of soft fruits. The larvae are usually present year-round in well-established compost heaps and walk on their backs. Photo by C.L. Hogue.

pillar hunters *(Calosoma)* may be kept together and do well on a diet of mealworms and crickets. Tiger beetles *(Cicindela)* can be kept for several weeks in a terrarium supplied with several centimeters of sand. Keep at least one corner of the tank moist and cover it with a light hood with a 40-watt aquarium bulb. Feed them every other day with a variety of live insects, worms, and crickets. Live food should be no larger than the beetles themselves.

Aquatic beetles are relatively easy to keep and provide hours of great beetle watching. Aquariums fitted with undergravel filters are a good system for beginners. After assembling the filter system, place a few centimeters of sealed aquarium gravel on top of the filter and fill the tank half-full with distilled or filtered water. Then add artificial or real plants and a few larger rocks so that your beetles have a place to rest or hide when submerged, then top off the water level of the tank.

Adding branches to the tank may be attractive but may also discolor the water. A light hood is essential not only for illuminating your aquarium but also to prevent your beetles from escaping. Most aquatic beetles such as whirligigs and predaceous diving beetles eat living or dead crickets placed on the surface of the water. Mosquito and mayfly larvae are excellent sources of food if they are available in sufficient quantities. Bits of raw meat are readily accepted by hungry predators and scavengers alike, but quickly foul the water. Water scavenger beetles (Hydrophilidae) feed on bits of lettuce or algae. Living aquarium plants or submerged rocks covered with algae provide food for most herbivorous species.

Pet Shop Beetles

Bait shops and pet stores often carry mealworms (Tenebrionidae) as fish bait or as food for pet birds, fish, amphibians, and reptiles. Entomological laboratories also raise mealworms by the millions, using them to test the efficacy of insecticides and for other experimental purposes. Anglers are well acquainted with using mealworms as bait. They are also used as live food in outdoor bird feeders.

The Common Mealworm, or Yellow Mealworm *(Tenebrio molitor)* (adult length 12 to 18 mm) (pl. 50), is originally from Europe. There, it sometimes becomes a pest of stored grain and grain products in mills and warehouses. Both adult and larval mealworms are nocturnal, inhabiting undisturbed accumulations of grain in dark corners or beneath bags of feed, but seldom become household pests. Mealworms also attack damp grain stored in bins.

The elongate beetles are dark brown to black with distinctly grooved elytra. The female lays 300 to 1,000 bean-shaped eggs directly on the food medium, either singly or in small clusters. The naturally sticky eggs soon become covered with bits of food, hiding them from potential scavengers and

Plate 50. Larvae of the Common Mealworm, or Yellow Mealworm (*Tenebrio molitor,* Tenebrionidae), are commonly sold in pet shops as live food for birds, fish, reptiles, and amphibians. Colonies of mealworms are easy to keep in the laboratory, classroom, or nature center and require little maintenance.

predators. In about two weeks the eggs hatch into slender, cylindrical, pale larvae. Young mealworms quickly darken, assuming their characteristic yellowish hues. The larval stage lasts approximately three months, during which time they molt 14 or 15 times. The mature larva measures approximately 25 mm in length and is yellowish overall but is more yellowish to brown toward the ends and along the margins of the body segments. The larvae prefer to pupate in secluded locations. The soft, white pupae are quite vulnerable and are protected only by their abdominal *gin-traps,* toothlike structures located on opposing abdominal segments that snap together to discourage predators and parasites. The pupal stage lasts about two weeks. Adults generally emerge in spring and early summer and live for two or three months. Typically, one full generation of mealworms is produced per year. Depending upon temperature conditions and the availability of food, mealworms can develop in as little as four months or may take as long as four years to complete their life cycle.

Pet stores and bait shops also sell giant and king mealworms, which are nothing more than common mealworms pumped up on juvenile hormones, compounds that prevent maturation. Chemically engineered to never grow up, these larvae never pupate but simply continue to eat and grow. Although fine for use as fishing bait, these hormonally enhanced larvae are not recommended for use as food for other animals.

Raising Mealworms

Colonies of mealworms are easy to keep in the laboratory, classroom, or nature center and require little maintenance. They are housed in a variety of broad, flat containers made of glass, metal, or plastic. The container should be at least 10 cm (4 in.) deep and have a tight-fitting screened lid to prevent escapes and discourage infestations by other grain pests.

Cover the bottom of the container with a few centimeters of meal, such as wheat bran, cornmeal, chicken mash, or oatmeal. Supplements of fish flakes, dry poultry mash, or ground-up dried pet food added to the meal enhance the nutritional value of the larvae. Cover the medium with two pieces of clean burlap. The space between the layers of burlap serves as a relatively secluded pupation site for the larvae.

Never water the colony, but add fruits or vegetables weekly to provide sufficient moisture. Place thin slices of potato, apple, carrot, or oranges on a piece of paper towel or cardboard to avoid spoiling the meal. Replace uneaten vegetables regularly to prevent the development of harmful molds and mite infestations.

Mealworms develop rapidly if kept at temperatures between 75 and 80 degrees F. As their waste, or frass, accumulates, a slight ammonia odor is evident. Remove the frass and replace it with fresh meal or remove all stages of beetles by sifting and place them in a new container supplied with fresh meal and burlap.

Mealworm larvae can be kept refrigerated indefinitely at temperatures around 38 degrees F, but soon die if they are frozen. Temporary refrigeration is sometimes used as a means of staggering the emergence times of the colony.

Raising Super Mealworms

The Super Mealworm is the larva of a large, black darkling beetle, *Zophobas morio* (adult length 22 to 27 mm), a native of Central and South America. Like the Common Mealworm, the Super Mealworm is frequently sold as live food or fishing bait. There is much anecdotal information on the care of super mealworms, but only the biology and natural history of *Z. atrator,* a species commonly reared in the laboratory, appears in the scientific literature. The natural history and rearing information presented here is based on *Z. atrator* and is applicable to the closely related *Z. morio.*

Adult beetles and larvae are often found roaming about or burrowing through guano deposits in bat caves, compost, or other accumulations of organic debris. Adults are usually found on the surface of the medium and deposit their eggs in it. The larvae spend their lives tunneling within the medium, feeding as they go. Depending upon temperature and quality of food, the larvae take several months to mature.

If kept under crowded conditions throughout their lives, the larvae continue to grow but never pupate. Eventually they begin to lose their vitality and slowly die. Once they reach a certain minimum weight and age, however, isolated larvae pupate immediately. Under natural conditions, larvae ready to pupate escape the crowded food mass to undergo metamorphosis in an isolated site where they are not disturbed. This behavior probably evolved as an adaptation to protect helpless pupae from the cannibalistic ravages of other beetle larvae.

The Super Mealworm does very well on a diet of bran or other grain products. In fact, housing and care for developing the Super Mealworm is identical to that of the Common Mealworm. Be sure to add meal supplements to colonies used as a live food source for pets. Mature Super Mealworms are very sensitive to crowded situations, and metamorphosis is actually hindered by continual touching by other larvae. When ready to pupate, larvae 3 to 4 cm (1.5 in.) or longer typically wander around the edges of their container, searching for an isolated site to pupate. Pupation is induced by carefully placing these individuals in separate black film canisters or some other small, dark enclosure supplied with a pinch or two of grain. Gently check the containers every week or so and return any adults to the colony so they can mate and lay eggs.

Beetles as Educational Tools

Live beetles are easily kept in the home, classroom, museum, or nature center. They are readily obtained in both urban and rural environments. In terms of management and care, beetles require little space, are inexpensive to maintain, and are easily displayed. They make great educational tools. Caring for them in the classroom instills a sense of responsibility among the students, helping them to develop the awareness that beetles, just like humans, have basic environmental and nutritional needs that must be met in order for them to thrive.

Observing Captive Beetles

Simple experiments with beetles can reveal the effects of temperature on the rate of their development, food preferences, and a variety of other interesting behaviors. One of the most effective, yet least understood, elements of beetle-based education is the power of touch. To talk about the defensive strategies of darkling beetles is one thing, but placing a live desert ironclad beetle *(Asbolus)* in a child's hand to demonstrate how it feigns death to escape the notice of predators is an experience that will not be soon forgotten. These and other beetle behaviors are ready subjects for investigation that can be explored by means of the scientific method and developed for local, county, and state science fairs. Regional surveys and mark and recapture studies, coordinated with researchers at universities and museums, are easily managed student projects that provide real data useful to the scientific community and lifelong memories for teachers and students alike.

Beetle Walks

If you are fortunate enough to live or work near a park or vacant lot, consider taking a small group out to observe beetles and other insects in nature. Always survey the area in advance to locate beetle hot spots, guaranteeing the participants plenty of opportunities to watch beetles up close. Be sure to observe local regulations and obtain necessary permits or permission in advance. Always have a backup date in case of bad weather. Look for as many examples as possible of different beetle families. Rather than lecturing participants on the identity of what they see, help them to discover the ecological roles occupied by beetles and other insects living in the habitat. Keep the group size small (20 or fewer) so everyone can see and participate. Encourage participants to ask questions and share their own discoveries with the group.

Photographing Beetles

Beetle photography is another method of collecting and can be as satisfying, if not more so, than making a specimen collection. Your photos can be used to document distribution and record behavior, or they may be used in scientific and popular publications. Civic, environmental, and educational organizations are always looking for speakers, and your images can dramatically illustrate lectures and workshops for these groups.

The expense and technicalities of insect photography are no doubt intimidating to some. Camera bodies, lenses, and flashes can be expensive, but good results can be obtained from moderately priced systems. With a bit of knowledge, some patience, and a bit of luck, it is possible for the beginning beetle photographer to achieve encouraging results from the start. The camera technology of today is such that the beetle photographer can be more of an artist and less of a technician. Remember that there is no single way to photograph insects and that every photographer has his or her favorite setup and method of working. Quality of light, film, camera, and lenses all come down to personal choice and budget. Nevertheless, a basic background in close-up photography or macrophotography will offer a solid foundation from which you can begin to experiment and develop your own techniques and style in photographing beetles and other insects.

Choosing your camera system is a long-term proposition and should not be entered into lightly. First, select the camera body. A single-lens-reflex (SLR) camera is essential, allowing you to accurately compose a photograph by viewing the scene directly through the lens rather than through a viewfinder. The automatic systems of most modern SLR cameras eliminate the guesswork required with older models, but reliance on these features seldom results in sharp images of beetles and other insects. Fortunately, you can override the automatic feature, selecting instead a combination of shutter speed and aperture that ensures a properly exposed, razor-sharp image. A properly exposed photograph means that the colors are not too bright or washed out, nor are they too dark or muddy.

Most SLR cameras come with a 50 mm lens, which is useful for taking pictures of people and landscapes. These lenses do not allow you to focus closely enough for insect photography, however. A 90 or 100 mm macrolens allows you to focus on beetles at relatively short distances. This close-focusing capability allows you to fill the frame of your photograph with the beetle's image or obtain a life-size image of the beetle on the film. The life-size image capability is often referred to as the ratio 1:1. There are several ways of achieving close focus capability, but the simplest is to use a 90 or 100 mm macrolens. Macrolenses with longer focal lengths of 150 or 200 mm allow you to be further away from your subject and still offer a 1:1 capability, but they are bulky and very expensive.

Although there are many types of film on the market, nothing beats the clarity, affordability, and flexibility of color slide film. Slides are cheaper to develop than prints and, if need be, it is easy to make a print from a slide. Prints may be easily shared one-to-one, but slides can be shared with an audience. If you are planning to publish your slides in books and magazines, editors prefer working with slides. Film sensitivity (i.e., the amount of light required for proper exposure) is measured in terms of an International Standards Organization (ISO) number. Film with an ISO number of 200 is much more sensitive to light and is said to be faster than film rated at ISO 25. In other words, ISO 200 film requires less light to expose it properly and can be used more effectively under low-light conditions. So why use slower film? Slower films are often more desirable, especially when photographing beetles, because of their richer colors, truer color replication, and sharper, less grainy images. Professional films rated between ISO 50 and 100 are usually best.

In addition to film sensitivity, proper exposure is controlled by two additional elements: the aperture of the camera lens, or f-stop, and the length of time (shutter speed) the camera exposes the film to the light. Another critical component of beetle photography is the depth of focus in a picture, usually referred to as depth of field. Depth of field is determined by the aperture of the iris in the lens and increases as the f-stop increases. As the f-stop increases, however, the amount of light reaching the film is reduced. Shooting at a slower shutter speed increases the amount of light reaching the film but can also reduce the sharpness of the picture as a result of movement on behalf of the camera or the subject. In order to use slower film in conjunction with smaller apertures and faster shutter speeds, you have to add more light to the subject by using flash equipment.

Through-the-lens (TLL) flash systems make flash photography easy. The long lens used in macrophotography requires placing one or more flashes near the end of the lens barrel using an off-camera flash bracket. Using the camera's own hot shoe for macrophotography places the flash too far back from the end of the lens. The lens then casts a shadow across your scene, ruining the picture. With a single flash in place at a 30 degree angle over the lens barrel to create the effect of natural morning or afternoon sunlight, a shutter speed of 1/125 of a second at f/16 or higher, you begin to achieve the desired results. As your photography improves you can begin experimenting with different flash positions and backgrounds to vary your results. All of the beetles in this book were photographed with one or two flashes at f/16 or f/22.

When you first start out, shoot a lot of film. Try different camera settings and take careful notes. By carefully comparing your results and notes you soon establish what settings work best for your camera under specific conditions. If you label your slides with the same locality data used for beetle specimens, your images also become part of a permanent record of your travels and observations. As important records you should consider storing your slides in an archival storage system that limits their long-term exposure to light, dust, and chemicals.

Digital photography is advancing quickly, allowing photographers to instantly check their results and discard unwanted images. Images taken with digital cameras are easily downloaded on computers and are easily e-mailed or used on personal Web sites. As of this writing, digital cameras equipped with suitable macrolenses are still prohibitively expensive. In time, this will probably change. For now, stick with using film, especially if you are interested in submitting images for publication.

Watching Beetles with Binoculars

Birders are not the only naturalists to effectively use binoculars. Many people are now watching butterflies and dragonflies through binoculars. Many beetles, especially conspicuous flower-feeding and predatory species, even those living in waters with a still surface, can be easily observed through binoculars. Binoculars with close-focusing capability that allow you to focus on scenes 2 m (6 ft) away or less are ideal for observing the details of beetle behavior. Even though many beetles are easily approached, viewing their magnified images through binoculars can be awe inspiring. Close-focusing monoculars are also very effective, inexpensive, and easy to pack in a collecting bag.

If you already have a pair of compact binoculars of the reverse-porro-prism design, you can modify them for close-up beetle watching. These binoculars have their front lenses set closer together than the eyepiece lenses. Purchase a two-element Nikon 5T close-up lens and affix it to the front of your binoculars with a soft lens hood and rubber bands to create close-focusing capability.

CHECKLIST OF NORTH AMERICAN BEETLE FAMILIES

Note: After Arnett and Thomas, 2001; Arnett, et al., 2002. Families with an asterisk * are not known to be established in California.

Suborder Archostemata

- ☐ Reticulated beetles (Cupedidae)
- ☐ Telephone-pole beetles (Micromalthidae)*

Suborder Myxophaga

- ☐ Minute bog beetles (Microsporidae)
- ☐ Skiff beetles (Hydroscaphidae)

Suborder Adephaga

- ☐ Wrinkled bark beetles (Rhysodidae)
- ☐ Ground and tiger beetles (Carabidae)
- ☐ Whirligig beetles (Gyrinidae)
- ☐ Crawling water beetles (Haliplidae)
- ☐ False ground beetles (Trachypachidae)
- ☐ Burrowing water beetles (Noteridae)
- ☐ Trout-stream beetles (Amphizoidae)
- ☐ Predacious diving beetles (Dytiscidae)

Suborder Polyphaga
Series Staphyliniformia
Superfamily Hydrophiloidea

- ☐ Water scavenger beetles (Hydrophilidae)
- ☐ False clown beetles (Sphaeritidae)
- ☐ Hister beetles, or clown beetles (Histeridae)

Superfamily Staphylinoidea

- ☐ Minute moss beetles (Hydraenidae)
- ☐ Feather-winged beetles (Ptiliidae)
- ☐ Primitive carrion beetles (Agyrtidae)
- ☐ Round fungus beetles (Leiodidae)
- ☐ Antlike stone beetles (Scydmaenidae)
- ☐ Carrion beetles (Silphidae)
- ☐ Rove beetles (Staphylinidae)

Series Scarabaeiformia
Superfamily Scarabaeoidea (formerly Lamellicornia)

- ☐ Stag beetles (Lucanidae)
- ☐ False stag beetles (Diphyllostomatidae)
- ☐ Bess beetles (Passalidae)*
- ☐ Enigmatic scarab beetles (Glaresidae)
- ☐ Hide beetles (Trogidae)
- ☐ Rain beetles (Pleocomidae)
- ☐ Earth-boring scarab beetles (Geotrupidae)
- ☐ Sand-loving scarab beetles (Ochodaeidae)
- ☐ Scavenger scarab beetles (Hybosoridae)
- ☐ Pill scarab beetles (Ceratocanthidae)*
- ☐ Bumblebee scarab beetles (Glaphyridae)
- ☐ Scarab, dung, May, June beetles, chafers (Scarabaeidae)

Series Elateriformia (formerly Dascilliformia; includes Eucinetiformia)
Superfamily Scirtoidea

- ☐ Plate-thigh beetles (Eucinetidae)
- ☐ Minute beetles (Clambidae)
- ☐ Marsh beetles (Scirtidae)

Superfamily Dascilloidea

- ☐ Soft-bodied plant beetles (Dascillidae)
- ☐ Cedar beetles or cicada parasite beetles (Rhipiceridae)

Superfamily Buprestoidea

- ☐ Schizopodid or false jewel beetles (Schizopodidae)
- ☐ Metallic wood-boring or jewel beetles (Buprestidae)

Superfamily Byrrhoidea

- ☐ Pill beetles or moss beetles (Byrrhidae)
- ☐ Riffle beetles (Elmidae)
- ☐ Long-toed water beetles (Dryopidae)
- ☐ Travertine beetles (Lutrochidae)*
- ☐ Minute marsh-loving beetles (Limnichidae)
- ☐ Variegated mud-loving beetles (Heteroceridae)
- ☐ Water penny beetles (Psephenidae)
- ☐ Ptilodactylid beetles (Ptilodactylidae)
- ☐ Chelonariid beetles (Chelonariidae)*
- ☐ Forest stream beetles (Eulichadidae)
- ☐ Callirhipid beetles (Callirhipidae)*

Superfamily Elateroidea

- ☐ Artematopodidae beetles (Artematopodidae)
- ☐ Texas beetles (Brachypsectridae)
- ☐ Rare click beetles (Cerophytidae)
- ☐ False click beetles (Eucnemidae)
- ☐ Throscid beetles (Throscidae)
- ☐ Click beetles (Elateridae)
- ☐ Net-winged beetles (Lycidae)
- ☐ Long-lipped beetles (Telegeusidae)*
- ☐ Glowworms (Phengodidae)
- ☐ Fireflies, lightningbugs, glowworms (Lampyridae)
- ☐ False soldier and false firefly beetles (Omethidae)
- ☐ Soldier beetles (Cantharidae)

Series Bostrichiformia
Superfamily Derodontoidea

- ☐ Jacobsoniid beetles (Jacobsoniidae)*
- ☐ Tooth-neck fungus beetles (Derodontidae)

Superfamily Bostrichoidea (includes Dermestoidea)

- ☐ Nosodendrid beetles (Nosodendridae)
- ☐ Skin beetles (Dermestidae)
- ☐ Bostrichid beetles (Bostrichidae)
- ☐ Death watch and spider beetles (Anobiidae)

Series Cucujiformia
Superfamily Lymexyloniodea

☐ Ship-timber beetles (Lymexylidae)

Superfamily Cleroidea

☐ Bark-gnawing beetles, cadelles (Trogossitidae)
☐ Checkered beetles (Cleridae)
☐ Soft-winged flower beetles (Melyridae)

Superfamily Cucujoidea (formerly Clavicornia)

☐ Cryptic slime mold beetles (Sphindidae)
☐ Short-winged flower beetles (Brachypteridae)
☐ Sap beetles (Nitidulidae)
☐ Palmetto beetles (Smicripidae)
☐ Root-eating beetles (Monotomidae)
☐ Silvanid flat bark beetles (Silvanidae)
☐ Parasitic flat bark beetles (Passandridae)*
☐ Flat bark beetles (Cucujidae)
☐ Lined flat bark beetles (Laemophloeidae)
☐ Shining flower and shining mold beetles (Phalacridae)
☐ Silken fungus beetles (Cryptophagidae)
☐ Lizard beetles (Languriidae)
☐ Pleasing fungus beetles (Erotylidae)
☐ Fruitworms (Byturidae)
☐ False skin beetles (Biphyllidae)
☐ Bothriderid beetles (Bothrideridae)
☐ Minute bark beetles (Cerylonidae)
☐ Handsome fungus beetles (Endomychidae)
☐ Lady beetles (Coccinellidae)
☐ Minute hooded, minute fungus, and hooded beetles (Corylophidae)
☐ Minute brown scavenger beetles (Latridiidae)

Superfamily Tenebrionoidea

☐ Hairy fungus beetles (Mycetophagidae)
☐ Archaeocryptic beetles (Archeocrypticidae)*
☐ Minute tree-fungus beetles (Ciidae)

- ☐ Polypore fungus beetles (Tetratomidae)
- ☐ False darkling beetles (Melandryidae)
- ☐ Tumbling flower beetles (Mordellidae)
- ☐ Ripiphorid beetles (Ripiphoridae)
- ☐ Cylindrical bark beetles (Colydiidae)†
- ☐ Monommatid beetles (Monommatidae)†
- ☐ Zopherid beetles (Zopheridae)
- ☐ Darkling beetles (Tenebrionidae)
- ☐ Jugular-horned beetles (Prostomidae)
- ☐ Synchroa bark beetles (Synchroidae)*
- ☐ False blister beetles (Oedemeridae)
- ☐ False longhorn beetles (Stenotrachelidae)
- ☐ Blister beetles (Meloidae)
- ☐ Palm and flower beetles (Mycteridae)
- ☐ Conifer bark beetles (Boridae)
- ☐ Dead log beetles (Pythidae)
- ☐ Fire-colored beetles (Pyrochroidae)
- ☐ Narrow-waisted bark beetles (Salpingidae)
- ☐ Antlike flower beetles (Anthicidae)
- ☐ Antlike leaf beetles (Aderidae)
- ☐ False flower beetles (Scraptiidae)

Superfamily Chrysomeloidea

- ☐ Longhorn beetles (Cerambycidae)
- ☐ Bean weevils (Bruchidae)
- ☐ Megalopodid leaf weevils (Megalopodidae)
- ☐ Orsodacnid leaf beetles (Orsodacnidae)
- ☐ Leaf beetles (Chrysomelidae)

Superfamily Curculionoidea

- ☐ Pine flower snout beetles (Nemonychidae)
- ☐ Fungus weevils (Anthribidae)
- ☐ Cycad weevils (Belidae)*

†Considered by some entomologists to be a subfamily of Zopheridae.

Superfamily Curculionoidea (continued)

- ☐ Leaf rolling and thief weevils, tooth-nose snout beetles (Attelabidae)
- ☐ Straight-snouted and pear-shaped weevils (Brentidae)
- ☐ New York weevils (Ithyceridae)*
- ☐ Weevils or snout beetles (Curculionidae)

CALIFORNIA'S SENSITIVE BEETLES

The following four species are listed on the Federal Endangered Species List:

Ground Beetles and Tiger Beetles (Carabidae)

Ohlone Tiger Beetle *(Cicindela ohlone)*, endangered
Delta Green Ground Beetle *(Elaphrus viridis)*, threatened

June Beetles (Scarabaeidae)

Mount Hermon June beetle *(Polyphylla barbata)*, endangered

Longhorn Beetles (Cerambycidae)

Valley Elderberry Longhorn Beetle *(Desmocerus californicus dimorphus)*, threatened

The following species live in sensitive habitats and many of them have been proposed for listing on the Federal Endangered Species List as threatened or endangered species:

Ground Beetles and Tiger Beetles (Carabidae)

Sacramento Valley Tiger Beetle *(Cicindela hirticollis abrupta)*
Sandy Beach Tiger Beetle *(Cicindela hirticollis gravida)*
Oblivious Tiger Beetle *(Cicindela latesignata obliviosa)*, extinct?
Greenest Tiger Beetle *(Cicindela tranquebarica viridissima)*
San Joaquin Tiger Beetle (*Cicindela tranquebarica* undescribed subspecies)
South Forks Ground Beetle *(Nebria darlingtoni)*
Siskiyou Ground Beetle *(Nebria gebleri siskiyouensis)*

Trinity Alps Ground Beetle *(Nebria sahlberbii triad)*
Scaphinotus behrensi
Humboldt Ground Beetle *(Scaphinotus longiceps)*

Predaceous Diving Beetles (Dytiscidae)

Death Valley Agabus Diving Beetle *(Agabus rumppi)*
Wooly Hydroporus Diving Beetle *(Hydroporus hirsutus)*
Leech's Skyline Diving Beetle *(Hydroporus leechi)*
Simple Hydroporus Diving Beetle *(Hydroporus simplex)*
Curved-footed Hygrotus Diving Beetle *(Hygrotus curvipes)*
Travertine Band-thigh Diving Beetle *(Hygrotus fontinalis)*

Water Scavenger Beetles (Hydrophilidae)

Leech's Chaetarthrian Water Scavenger Beetle *(Chaetarthria leechi)*
Ricksecker's Water Scavenger Beetle *(Hydrochara rickseckeri)*

Minute Moss Beetles (Hydraenidae)

Wing-shoulder Minute Moss Beetle *(Ochthebius crassalus)*
Wilber Springs Minute Moss Beetle *(Ochthebius reticulatus)*

Enigmatic Scarab Beetles (Glaresidae)

Kelso Dune Glaresis Scarab *(Glaresis arenata)*

Rain Beetles (Pleocomidae)

Santa Cruz Rain Beetle *(Pleocoma conjugens conjugens)*

Bumblebee Scarabs (Glaphyridae)

White Sand Bear Scarab *(Lichnanthe albopilosa)*
Bumblebee Scarab *(Lichnanthe ursina)*

Scarab Beetles and June Beetles (Scarabaeidae)

Ciervo Aegialian Scarab *(Aegialia concinna)*
Carlson's Dune Beetle *(Anomala carlsoni)*
Hardy's Dune Beetle *(Anomala hardyorum)*
San Clemente Island Coenonycha Beetle *(Coenonycha clementina)*
Saline Valley Snow-front June Beetle *(Polyphylla anteronivea)*
Death Valley June Beetle *(Polyphylla erratica)*
Atascadero June Beetle *(Polyphylla nubila)*

Delta June Beetle *(Polyphylla stellata)*
Andrews' Dune Scarab *(Pseudocotalpa andrewsi)*

Riffle Beetles (Elmidae)

Wawona Riffle Beetle *(Atractelmis wawona)*
Brownish Dubiraphian Riffle Beetle *(Dubiraphia brunnescens)*
Giuliani's Dubiraphian Riffle Beetle *(Dubiraphia giulianii)*
Microcylloepus formcoideus
Microcylloepus similis
Pinnacles Optioservus Riffle Beetle *(Optioservus canus)*

False Click Beetles (Eucnemidae)

Dohrn's Elegant Eucnemid Beetle *(Paleoxenus dohrni)*

Darkling Beetles (Tenebrionidae)

Globose Dune Beetle *(Coelus globosus)*
San Joaquin Dune Beetle *(Coelus gracilis)*
Channel Islands Dune Beetle *(Coelus pacificus)*

Blister Beetles (Meloidae)

Hopping's Blister Beetle *(Lytta hoppingi)*
Mojave Desert Blister Beetle *(Lytta insperata)*
Moestan Blister Beetle *(Lytta moesta)*
Molestan Blister Beetle *(Lytta molesta)*
Morrison's Blister Beetle *(Lytta morrisoni)*

Antlike Flower Beetles (Anthicidae)

Antioch Dunes Anthicid *(Anthicus antiochensis)*
Sacramento Anthicid *(Anthicus sacramento)*

Longhorn Beetles (Cerambycidae)

Rude's Longhorn Beetle *(Necydalis rudei)*

Weevils (Curculionidae)

Nelson's Miloderes Weevil *(Miloderes nelsoni)*
Lange's El Segundo Dune Weevil *(Onychobaris langei)*
Blaisdell's Trigonoscuta Weevil *(Trigonoscuta blaisdelli)*
Brown-tassel Trigonoscuta Weevil *(Trigonoscuta brunnotasselata)*

Weevils (continued)

Santa Catalina Island Trigonoscuta Weevil *(Trigonoscuta catalina)*
Dorothy's El Segundo Dune Weevil *(Trigonoscuta dorothea dorothea)*
Doyen's Trigonoscuta Dune Weevil *(Trigonoscuta doyeni)*
Santa Cruz Island Shore Weevil *(Trigonoscuta stantoni)*

COLLECTIONS, SOCIETIES, AND OTHER RESOURCES

Some Beetle Collections in California

Essig Museum of Entomology, Department of Entomological Sciences, University of California, Berkeley, CA 94720

The Bohart Museum of Entomology, University of California, Davis, CA 95616

Death Valley Museum, Death Valley National Monument, Death Valley, CA 92328

Entomology Section, Natural History Museum of Los Angeles County, 900 Exposition Blvd., Los Angeles, CA 90007

Museum Collections of Lassen Volcanic Park, Lassen Park, Mineral, CA 96063

Insect Collection, Department of Biology, California State University, Northridge, CA 91330

Museum Collections of Pinnacles National Monument, Paicines, CA 95043

Diablo Valley College, 321 Golf Club Rd., Pleasant Hill, CA 94533

Entomology Department, California Polytechnic University, 3801 Temple Ave., Pomona, CA 91768

Entomology Museum, University of California, Riverside, CA 92521

California Department of Food and Agriculture-Plant Pest Diagnostics Center, 3294 Meadowview Rd., Sacramento, CA 95832–1448

Department of Invertebrate Zoology, Santa Barbara Museum of Natural History, 2559 Puesta Del Sol Rd., Santa Barbara, CA 93105

Entomology Department, San Diego Natural History Museum, P.O. Box 1390, San Diego, CA 92112

Department of Entomology, California Academy of Sciences, Golden Gate Park, San Francisco, CA 94118

J. Gordon Edwards Museum of Entomology, Department of Biological Sciences, San Jose State University, One Washington Square, San Jose, CA 95192–0100

Museum Collections of Joshua Tree National Monument, 74485 National Monument Dr., Twentynine Palms, CA 92277

Societies and Web Sites Promoting the Study of Beetles

The California Beetle Project is sponsored by the Santa Barbara Museum of Natural History. It will produce an up-to-date inventory of California beetles and eventually compile distributional information by county for all beetle species occurring in the state, provide identification resources, including keys and images for the most prominent species, and compile a complete bibliography of literature pertaining to the taxonomy and natural history of California beetles. The project is designed to spur interest and to facilitate additional revisionary research on California beetles and encourage broader evolutionary studies using Coleoptera, particularly toward understanding California biogeography. www.sbnature.org/collections/invert/entom/cbphomepage.html.

The Coleopterists' Society, c/o Terry N. Seeno, Treasurer, CDFA-Plant Pest Diagnostics Center, 3294 Meadowview Rd., Sacramento, CA 95832–1448. Publishes the *Coleopterists' Bulletin,* an international journal devoted to the study of beetles, including many scientific articles on California species. www.coleopsoc.org.

The Lorquin Entomological Society, c/o Entomology Section, Natural History Museum of Los Angeles County, 900 Exposition Blvd., Los Angeles, CA 90007. www.nhm.org/research/entomology/LorquinSoc.

The Pacific Coast Entomological Society, c/o Department of Entomology, California Academy of Sciences, Golden Gate Park, San Francisco, CA 94118. Also publishes the *Pan-Pacific Entomologist,* a treasure trove of taxonomic and natural history information on California beetles. www.solpugid.com/PCES/pceshome.html.

The Young Entomologists Society, 6907 West Grand River Ave., Lansing, MI 48906. Publishes various newsletters and guides for the beginning insect enthusiast. members.aol.com/YESbugs/bugclub.html.

California Sources for Books and Collecting Equipment

Acorn Naturalists, 155 El Camino Real, Tustin, CA 92780; phone (800) 422-8886; e-mail Acorn@aol.com; www.acornnaturalists.com.

Specializing in resources for science and environmental education through investigation and experimentation.

BioQuip Products, 2321 Gladwick St., Rancho Dominguez, CA 90220; phone (310) 667-8800. The most comprehensive single source of entomological supplies and references materials. E-mail bioquip @aol.com; www.bioquip.com.

SELECTED REFERENCES

Chapter 1. A Brief History of Beetle Study in California

Caltagirone, L.E., and R.L. Doutt. 1989. The history of the vedalia beetle importation to California and its impact on the development of biological control. *Annual Review of Entomology* 34:1–16.

Calvert, P.P. 1898. A biographical notice of George Henry Horn. *Transactions of the American Entomological Society* 25:i–xxiv.

Chemsak, J. 2000. Earle Gorton Linsley. *American Entomologist* 46:270–71.

Dow, R.P. 1914. The Russian masters in Coleoptera. *Bulletin of the Brooklyn Entomological Society* 9:96–101.

Essig, E.O. 1927. Some insects from the adobe walls of the old missions of lower California. *Pan-Pacific Entomologist* 3 (4): 194–95.

———. 1931. *A history of entomology.* New York: MacMillan Company. 1029 pp.

———. 1934. The historical background of entomology in relation to the early development of agriculture in California. *Pan-Pacific Entomologist* 10 (1): 1–11; 10 (2): 49–58; 10 (3): 97–101.

———. 1953. Edwin Cooper Van Dyke. *Pan-Pacific Entomologist* 29 (2): 73–97.

Grinnell, F. 1914. The development of California entomology. *Bulletin of the Brooklyn Entomological Society* 9 (4): 67–73.

Hatch, M.H. 1926. Thomas Lincoln Casey as a coleopterist. *Entomological News* 37 (6): 175–79.

Herman, L.H. 2001. Catalog of the Staphylinidae (Insecta: Coleoptera). 1758 to the end of the second millennium. I. Introduction, history, biographical sketches and omaliine group. *Bulletin of the American Museum of Natural History* 265:1–650.

Kavanaugh, D.H., and P.H. Arnaud, Jr. 1981. Hugh Bosdin Leech—A curator's curator. *Pan-Pacific Entomologist* 57 (1): 2–42.

Leech, R. 1991. Hugh Bosdin Leech (1910–1990). *Coleopterists' Bulletin* 45 (1): 95–96.

Leng, C.W. 1925. Thomas Lincoln Casey. *Entomological News* 36 (1): 97–100.

Linsley, E.G. 1940. Henry Clinton Fall. *Pan-Pacific Entomologist* 16 (1): 1–3.

Linsley, E.G., ed. 1978. *Beetles from the early Russian explorations of the west coast of North America 1815–1857.* New York: Arno Press.

Mallis, A. 1971. *American entomologists.* New Brunswick, N.J.: Rutgers University Press. 549 pp.

Osborn, H. 1937. *Fragments of entomological history.* Columbus, Ohio: H. Osborn. 394 pp.

Scudder, S.H. 1884. A biographical sketch of Dr. John Lawrence LeConte. *Transactions of the American Entomological Society* 11:i–xxvii.

Skinner, H. 1898. Dr. George H. Horn. *Entomological News* 9:1–3.

Sorensen, W.C. 1995. *Brethren of the net: American entomology, 1840–1880.* Tuscaloosa, Ala.: The University of Alabama Press. 357 pp.

Van Dyke, E.C. 1947. The biography of Frank Ellsworth Blaisdell, Sr. *Pan-Pacific Entomologist* 23 (2): 49–58.

Chapter 2. Form, Diversity, and Classification

Brown, L., and L.L. Rockwood. 1986. On the dilemma of horns. *Natural History* 104 (7): 55–61.

Chapman, R.F. 1998. *The insects: Structure and function.* New York: Elsevier. 770 pp.

Crawford, C.S. 1981. *Biology of desert invertebrates.* Berlin: Springer-Verlag. 314 pp.

Crowson, R.A. 1981. *The biology of the Coleoptera.* London: Academic Press. 802 pp.

Eberhard, W.G. 1980. Horned beetles. *Scientific American* 242 (3): 166–82.

Emlen, D.J. 2000. Integrating development with evolution: a case study with beetle horns. *Bioscience* 50 (5): 403–18.

Farrell, B.D. 1998. "Inordinate fondness" explained: why are there so many beetles? *Science* 281:555–59.

Hadley, N.F. 1993. Beetles make their own waxy sunblock. *Natural History* 102 (8): 44–45.

Stehr, F.W., ed. 1991. *Immature insects.* Vol. 2. Dubuque, Iowa: Kendall/Hunt Publishing Company. 975 pp.

Chapter 3. The Lives of Beetles

Alcock, J. 1994. Postinsemination associations between males and females in insects: The mate-guarding hypothesis. *Annual Review of Entomology* 39:1–21.

Evans, D.L., and J.O. Schmidt, eds. 1990. *Insect defenses: Adaptive mechanisms and strategies of prey and predators.* Albany, N.Y.: State University of New York Press. 482 pp.

Hoagland, M., and B. Dodson. 1995. *The way life works.* New York: Times Books, Random House. 233 pp.

Kistner, D.H. 1982. The social insect's bestiary. In Vol. 3, *Social insects,* ed. H.R. Hermann, 1–244. New York: Academic Press. 459 pp.

Matthews, R.W., and J.R. Matthews. 1978. *Insect behavior.* New York: John Wiley & Sons. 507 pp.

McIver, J.D., and G. Stonedahl. 1993. Myrmecomorphy: morphological and behavioral mimicry of ants. *Annual Review of Entomology* 38:351–79.

Milne, L.J., and M. Milne. 1976. The social behavior of burying beetles. *Scientific American* 235 (2): 84–89.

Murlis, J. 1992. Odor plumes and how insects use them. *Annual Review of Entomology* 37:505–32.

Rettenmeyer, C.W. 1970. Insect mimicry. *Annual Review of Entomology* 15:43–74.

Chapter 4. Distribution of California Beetles

Bakker, E. 1984. *An island called California: An ecological introduction to its natural communities.* Berkeley: University of California Press. 484 pp.

Downs, J.A., and D.H. Kavanaugh, eds. 1988. *Origins of the North American insect fauna.* Memoirs of the Entomological Society of Canada, no. 44, 168 pp.

Fall, H.C. 1897. A list of the Coleoptera of the southern California islands, with notes and descriptions of new species. *Canadian Entomologist* 29:233–44.

Fall, H.C., and A.C. Davis. 1934. The Coleoptera of Santa Cruz Island. *Canadian Entomologist* 66:143–44.

Linsley, E.G. 1961. The Cerambycidae of North America. Part I. Introduction. *University of California Publications in Entomology* 18:1–135.

Mani, M.S. 1968. *Ecology and biogeography of high altitude insects.* The Hague, The Netherlands: Dr. W. Junk N.V. Publishers.

Menke, A.S., and D.R. Miller, eds. 1981. *Entomology of the California Channel Islands: Proceedings of the first symposium.* Santa Barbara Natural History Museum. 169 pp.

Miller, S.E., and A.S. Menke. 1981. *Entomological bibliography of the California Islands,* Occasional Paper Number 11. Santa Barbara Museum of Natural History. 78 pp.

Morain, S.A. 1984. *Systematic and regional biogeography.* New York: Van Nostrand Reinhold Company. 334 pp.

Nagano, C.D., S.E. Miller, and C.L. Hogue. 1983. Castaways of California. *Terra* 21 (4): 23–26.

Papp, R.P. 1978. Ecology and habitat preferences of high altitude Coleoptera from the Sierra Nevada. *Pan-Pacific Entomologist* 54 (3): 161–72.

Schoenherr, A.A. 1992. *A natural history of California.* Berkeley: University of California Press. 772 pp.

Schoenherr, A.A., C.R. Feldmeth, and M.J. Emerson. 1999. *Natural history of the islands of California.* Berkeley: University of California Press. 491 pp.

Van Dyke, E.C. 1901. Observations upon the faunal regions of California from the standpoint of a coleopterist. *Journal of the New York Entomological Society* 9 (4): 197–99.

———. 1919. The distribution of insects in North America. *Annals of the Entomological Society of America* 12:1–12.

———. 1926. Certain peculiarities of the Coleopterous fauna of the Pacific Northwest. *Annals of the Entomological Society of America.* 26 (1): 1–12.

———. 1939–40. The origins and distribution of the Coleopterous insect fauna of North America. *Proceedings of the Pacific Science Congress* (64): 255–68.

Zimmerman, M.L. 1990. Coleoptera found in imported stored-food products entering southern California and Arizona between December 1984 through December 1987. *Coleopterists' Bulletin* 44 (2): 235–40.

Chapter 5. Beetles of Special Interest

Anderson, J.R., and E.C. Loomis. 1978. Exotic dung beetles in pasture and range land ecosystems. *California Agriculture* (February): 31–32.

Arnold, R.A. 1983. Biological studies of the delta green ground beetle, *Elaphrus viridis* Horn (Coleoptera: Carabidae) at Jepson Prairie Preserve in 1983. Unpublished report. San Francisco: The Nature Conservancy.

Bright, D.E., and R.W. Stark. 1973. *The bark and ambrosia beetles of California (Coleoptera: Scolytidae and Platypodidae).* Bulletin of the California Insect Survey 16, 169 pp.

Collinge, S.K., M. Holyoak, C.B. Barr, and J.T. Marty. 2001. Riparian habitat fragmentation and population persistence of the threatened valley elderberry longhorn beetle in central California. *Biological Conservation* 100:103–13.

Connell, W.A. 1977. A key to the *Carpophilus* sap beetles associated with stored foods in the United States. *Cooperative Plant Pest Report* 2:398–404.

Coope, G.R. 1970. Interpretations of Quaternary insect fossils. *Annual Review of Entomology* 15:97–120.

Dowell, R.V., and R. Gill. 1989. Exotic invertebrates and their effects on California. *Pan-Pacific Entomologist* 65 (2): 132–45.

Doyen, J.T., and S.E. Miller. 1980. Review of Pleistocene darkling ground beetles of the California asphalt deposits (Coleoptera: Tenebrionidae, Zopheridae). *Pan-Pacific Entomologist* 56 (1): 1–10.

Elias, S.A. 1994. *Quaternary insects and their environments.* Washington, D.C.: Smithsonian Institution Press. 284 pp.

Essig, E.O. 1931. *A history of entomology.* New York: MacMillan Company. 1029 pp.

Fleming, W.E. 1972. *Biology of the Japanese beetle.* United States Department of Agriculture Technical Bulletin 1449. Washington, D.C.: Agricultural Research Service. 129 pp.

Gammon, E.T. 1961. The Japanese beetle in California. *California Department of Food and Agriculture Bulletin* 50:221–35.

Goulet, H. 1983. The genera of Holarctic Elaphrini and species of *Elaphrus* Fabricius (Coleoptera: Carabidae): Classification, phylogeny, and zoogeography. *Quaestiones Entomologicae* 19:219–482.

Grinnell, F., Jr. 1908. Quaternary myriopods and insects of California. *Bulletin of the Department of Geology: University of California Publications* 5 (12): 207–15.

Hoebeke, E.R., and K. Beucke. 1997. Adventive *Onthophagus* (Coleoptera: Scarabaeidae) in North America: Geographical ranges, diagnoses, and new distributional records. *Entomological News* 108 (5): 345–62.

Howarth, F.G. 1991. Environmental impacts of classical biological control. *Annual Review of Entomology* 36:485–509.

Lingafelter, S.W., and E.R. Hoebeke. 2002. *Revision of* Anoplophora *(Coleoptera: Cerambycidae).* Washington, D.C.: Entomological Society of Washington. 236 pp.

McDavid, G.E. 1981. Days of vines and roses: The Japanese beetle in California. *Agrichemical Age* (August–September): 48–52.

Miller, S.E. 1997. Late Quaternary insects of Rancho La Brea, California, USA. *Quaternary Proceedings* (5): 185–91.

Miller, S.E., R.D. Gordon, and H.F. Howden. 1981. Reevaluation of Pleistocene scarab beetles from Rancho La Brea, California (Coleoptera: Scarabaeidae). *Proceedings of the Entomological Society of Washington* 83 (4): 625–30.

Miller, S.E., and S.B. Peck. 1979. Fossil carrion beetles of Pleistocene California asphalt deposits, with a synopsis of Holocene California Silphidae (Insecta: Coleoptera: Silphidae). *Transactions of the San Diego Society of Natural History* 19 (8): 85–106.

Miller, W.E., and F.P. Keen. 1960. *Biology and control of the western pine beetle.* Miscellaneous Publication 800. Washington, D.C.: United States Department of Agriculture. 381 pp.

Moore, I., and S.E. Miller. 1978. Fossil rove beetles from Pleistocene California asphalt deposits (Coleoptera: Staphylinidae). *Coleopterists' Bulletin* 32 (1): 37–39.

Nagano, C.D., S.E. Miller, and A.V. Morgan. 1983. Fossil tiger beetles (Coleoptera: Cicindelidae): Review and new Quaternary records. *Psyche* 89:339–46.

Okumura, G.I., and I.E. Savage. 1974. Nitidulid beetles most commonly found attacking dried fruits in California. *National Pest Control Operators News* 34:2–7.

Penrose, R.L. 1985. The eucalyptus borer, a pest new to California. *California Plant Pest and Disease Report* 4 (3): 80–84.

Pierce, W.D. 1960. Fossil arthropods of California. No. 23. Silicified insects in Miocene nodules from the Calico Mountains. *Bulletin of the Southern California Academy of Sciences* 59 (1): 40–49.

———. 1961. The growing importance of palaeoentomology. *Proceedings of the Entomological Society of Washington* 63 (3): 211–17.

Spilman, T.J. 1976. A new species of fossil *Ptinus* from fossil wood rat nests in California and Arizona (Coleoptera, Ptinidae), with a postscript on the definition of a fossil. *Coleopterists' Bulletin* 30 (3): 239–44.

Strong, K.G. 1970. Distribution and relative abundance of stored-product insects in California: A method of obtaining sample populations. *Journal of Economic Entomology* 63:591–96.

Struble, G.R., and R.C. Hall. 1955. *The California five-spined engraver, its biology and control.* United States Department of Agriculture Circulars 965. Washington, D.C.: United States Department of Agriculture. 21 pp.

Tschinkel, W. R. 1993. Crowding, maternal age, age at pupation, and life history of *Zophobas atratus* (Coleoptera: Tenebrionidae). *Annals of the Entomological Society of America* 86 (3): 278–97.
Tyndale-Biscoe, M. 1990. *Common dung beetles in pastures of southeastern Australia.* Melbourne, Australia: Commonwealth Scientific and Industrial Research Organization. 71 pp.
White, R. E. 1971. Key to the North American genera of Anobiidae, with phylogenetic and synonymic notes (Coleoptera). *Annals of the Entomological Society of America* 64 (1): 179–91.
———. 1982. *A catalog of the Coleoptera of America north of Mexico: Family Anobiidae.* Agricultural Handbook 529–70. Washington, D.C.: United States Department of Agriculture. 58 pp.
———. 1990. *Lasioderma haemorrhoidale* (Ill.) now established in Califorina, with biological data on *Lasioderma* species (Coleoptera: Anobiidae). *Coleopterists' Bulletin* 44 (3): 362–64.
Zimmerman, M. L. 1990. Coleoptera found in imported stored-food products entering southern California and Arizona between December 1984 through December 1987. *Coleopterists' Bulletin* 44 (2): 235–40.

Chapter 6. Common and Conspicuous Families of California Beetles

Arnett, R. H. Jr., and M. C. Thomas, eds. 2001. *American beetles.* Vol. 1, *Archostemata, Myxophaga, Adephaga, Polyphaga: Staphyliniformia.* Boca Raton, Fla.: CRC Press. 443 pp.
Arnett, R. H. Jr., M. C. Thomas, P. E. Skelley, and J. H. Frank, eds. 2002. *American beetles.* Vol. 2, *Polyphaga: Scarabaeoidea through Curculionidae.* Boca Raton, Fla.: CRC Press. 861 pp.
Booth, R. G., M. L. Cox, and R. B. Madge. 1990. *IIE Guides to insects of importance to man.* Vol. 3, *Coleoptera.* Oxon, U.K.: International Institute of Entomology. 384 pp.
Bousquet, Y., ed. 1991. *Checklist of beetles of Canada and Alaska.* Publication 1861/E. Ottawa: Research Branch, Agriculture Canada. 430 pp.
Ebeling, W. 1978. *Urban entomology.* Berkeley: University of California Division of Agricultural Sciences. 695 pp.
Essig, E. O. 1915. Injurious and beneficial insects of California. *Supplement to the Monthly Bulletin, State Commission of Horticulture* 4 (4): 1–541.
———. 1926. *Insects of western North America.* New York: The MacMillan Company. 1035 pp.

Furniss, R.L., and V.M. Carolin. 1977. *Western forest insects.* Miscellaneous Publication 1339. Washington, D.C.: U.S. Department of Agriculture, Forest Service. 654 pp.

Hatch, M.H. 1953. *The beetles of the Pacific Northwest. Part I: Introduction and Adephaga.* Vol. 16, University of Washington Publications in Biology. Seattle: University of Washington. 340 pp.

———. 1957. *The beetles of the Pacific Northwest. Part II. Staphyliniformia.* Vol. 16, University of Washington Publications in Biology. Seattle: University of Washington. 384 pp.

———. 1962. *The beetles of the Pacific Northwest. Part III. Pselaphidae and Diversicornia I.* Vol. 16, University of Washington Publications in Biology. Seattle: University of Washington. 503 pp.

———. 1965. *The beetles of the Pacific Northwest. Part IV. Macrodactyles, Palpicornes, and Heteromera.* Vol. 16, University of Washington Publications in Biology. Seattle: University of Washington. 268 pp.

———. 1971. *The beetles of the Pacific Northwest. Part V. Rhipiceroidea, Sternoxi, Phytophaga, Rhynchophora, and Lamellicornia.* Vol. 16, University of Washington Publications in Biology. Seattle: University of Washington. 662 pp.

Hogue, C.L. 1993. *Insects of the Los Angeles Basin.* Los Angeles: Natural History Museum of Los Angeles County. 446 pp.

Papp, C.S. 1984. *Introduction to North American beetles.* Sacramento, Calif.: Entomography Publications. 335 pp.

Powell, J.A., and C.L. Hogue. 1979. *California insects.* California Natural History Guides 44. Berkeley: University of California Press. 388 pp.

Ritcher, P.O. 1966. *White grubs and their allies: A study of North American scarabaeoid larvae.* Corvalis, Oreg.: Oregon State University Press. 219 pp.

Solomon, J.D. 1995. *Guide to the insect borers of North American broadleaf trees and shrubs.* Agricultural Handbook 706. Washington, D.C.: U.S. Department of Agriculture, Forest Service. 735 pp.

Stehr, F.W., ed. 1991. *Immature insects.* Vol. 2. Dubuque, Iowa: Kendall/Hunt Publishing Company. 975 pp.

Usinger, R.L., ed. 1956. *Aquatic insects of California.* Berkeley: University of California Press. 508 pp.

White, R.E. 1983. *A field guide to the beetles of North America.* The Peterson Field Guide Series. Boston: Houghton Mifflin Company. 368 pp.

Chapter 7. Studying Beetles

Acorn, J. 2001. *Tiger beetles of Alberta.* Edmonton: University of Alberta Press.

Allen, R. L. 1999. *Stalking the wild arthropod: The Lorquin Entomological Society's guide to photographing arthropods.* The Lorquin Entomological Society Special Publication Number 1.

Besuchet, C., D. H. Burkhardt, and I. Löbl. 1987. The "Winkler/Moczarski" collector as an efficient extractor for fungus and litter Coleoptera. *Coleopterists' Bulletin* 41:392–94.

Davies, P. H. 1998. *The complete guide to close-up and macro photography.* Devon, U.K.: David & Charles.

DeLorme Mapping Company. 1998. *Northern California Atlas and Gazetter.* Freeport, Maine: DeLorme Mapping Company.

DeLorme Mapping Company. 1998. *Southern and Central California Atlas and Gazetter.* Freeport, Maine: DeLorme Mapping Company.

Dussart, G. 1990. Mark-recapture experiments with freshwater organisms. *Journal of Biological Education* 25 (2): 116–18.

Gist, C. S., and Crossley, D. A., Jr. 1972. A method for quantifying pitfall trapping. *Environmental Entomology* 2:951–52.

Grissell, E. 2001. *Insects and gardens: In pursuit of a garden ecology.* Portland: Timber Press.

Hicks, P. 1997. *Photographing butterflies and other insects.* Kingston-upon-Thames, U.K.: Fountain Press Ltd.

Martin, J. E. H. 1977. *The insects and arachnids of Canada. Part 1. Collecting, preparing, and preserving insects, mites, and spiders.* Publication 1643. Ottawa: Research Branch, Canada Department of Agriculture.

Moore, I., J. Hlavac, and S. Frommer. 1976. Instant relaxing of insects. *Coleopterists' Bulletin* 30 (1): 99–100.

Newton, A., and S. B. Peck. 1975. Baited pitfall traps for beetles. *Coleopterists' Bulletin* 29:45–46.

O'Brien, C. W. 1973. Night collecting with a headlamp and sheet. *Coleopterists' Bulletin* 27 (3): 153.

Peck, S. B., and A. E. Davies. 1980. Collecting small beetles with large-area "window" traps. *Coleopterists' Bulletin* 34 (2): 237–39.

Perkins, P. P. 1980. Preventing distortion in preserved beetle pupae. *Coleopterists' Bulletin* 34 (3): 284.

Pyle, R. M. 1992. *Handbook for butterfly watchers.* Boston: Houghton Mifflin Company.

Suter, W. R. 1966. Techniques for the collection of microcoleoptera of the families Pselaphidae, Ptiliidae, and Scydmaenidae. *Coleopterists' Bulletin* 20 (2): 33–38.

Thomas, D. B., Jr., and E. L. Sleeper. 1977. The use of pitfall traps for estimating the abundance of arthropods, with special reference to the Tenebrionidae (Coleoptera). *Annals of the Entomological Society of America* 70:242–48.

van der Berghe, E. 1992. On pitfall trapping invertebrates. *Entomological News* 103:149–56.

West, L., and J. Ridl. 1994. *How to photograph insects and spiders.* Mechanicsburg, Penn.: Stackpole Books.

Wheeler, Q. D., and J. V. McHugh. 1987. A portable and convertible "Moczarski/Tullgren" extractor for fungus and litter Coleoptera. *Coleopterists' Bulletin* 41:9–12.

INDEX

Page references in **boldface type** refer to main discussions of families, genera, or species.

abdomen
  larval, 48–49
  morphology, 27, 42
  pupal, 51
Abies, as beetle habitat, 214
Acmaeodera, wings, 40
adenosine triphosphate (ATP), bioluminescence role, 67
Adenostoma fasciculatum, as beetle habitat, 214
Aderidae, 186
adult stage, 66
aerial nets, 221–222
African Goliath beetle, 162
Agabus, distribution patterns, 89
Agricultural Experiment Stations, University of California, 9
agricultural pests
  feeding habits of, 70–71
  introduced, 3–4, 94–95
  research, xvi, 2, 9
  See also biological controls; *specific pests*
Agrilus
  galls, 167
  hyperici, 201
Agrytidae (agrytid beetle family), 156
air sacs, morphology, 45
Alaus, 169
  melanops, 77, 170
alcohol, in storage vials, 231
Alleculinae, 141
allometry, 33
Altica, 203
Ambrosia Beetle, California Oak, 108
ambrosia beetle family, 184, **204–207**
  as forest pests, 107–108
  parental care, 70
"America's Entomological Declaration of Independence," 13
amphibians, as beetle predators, 197
Amphizoidae, 191
Anaphes nitens, as biological control agent, 122
angiosperms, coevolution with beetles, 53
Animalia kingdom, 56, 57
Año Nuevo Islands, 87
Anobiidae, 181
  courtship behavior, 68
  fossils, 102–103
  as household pests, 113–114
  introduction to California, 4
Anoplophora glabripennis, **126–128**
antennae
  of aquatic beetles, 46
  connection to nervous system, 43
  larval, 48
  morphology, 27, 35–37, 68
antennameres, 37
Anthicidae, 38, 186, 192, 203
Anthonomus grandis, 94
Anthrenus, 179
  flavipes, 116

Anthrenus (continued)
introduction to California, 4
verbasci, 82–83 (photo), 116–117
Anthribidae, 207
ants
as beetle habitat, 28–29, 73, 151
biological controls for, 72
mimicry of, 78
aorta, dorsal, 44–45
Aphididae (aphids), biological controls for, 133, 187
Aphodius
feeding habits, 128
as turf pests, 164
Aplastus, 171
aposematic coloration, 79
apple bugs or apple smellers. *See* whirligig beetle family
aquatic beetles. *See* water beetles
Archodontes
melanopus aridus, 91
melanopus melanopus, 91–92
melanopus serrulatus, 91–92
Arthropoda phylum, 56, 57
Asbolus
distribution patterns, 91
observing, 250
thanatosis, 75
verrucosus, 28
Asilidae, 160
asphalt deposits, 99–102
aspirators, 228–229
Ataenius, as turf pests, 164
Attagenus, 179
Attelabidae, 207
Avetianella longoi, as biological control agent, 120–121

bark beetle family, 184, **204–207**
biological controls for, 152
feeding habits, 72
as forest pests, 108–112
parental care, 70
stridulation, 68
Bark Beetle(s)
Cylindrical, 184
Flat, 29, 197
Minute, 192
Mountain, 112
Narrow-waisted, 141, 186
bats, as beetle predators, 197
beating sheets, 223
bees and wasps
as biological control agents, 120–121, 122
biological controls for, 74, 184, 193
mimics of, 77–78, 195
nests as beetle habitat, 180
beetle collection equipment
aspirators, 228–229
beating sheets, 223
Berlese funnels, 215, 223–224
brushes, 227–228
cameras, 250–254
forceps, 227–228
killing jars and agents, 229–230
nets, 221–222
night collecting, 224–226
strainers, 223
traps, 128, 226–227
beetle collections
best seasons for making, 210–214
California Academy of Sciences, 8–9
classifying, 56–57
of early explorers, 5–8, 14
ethical concerns, 218–221
holotypes, 55
labeling, 53–54, 237
larvae and pupae, preserving, 238–239
live specimens, 241–245
mounting and spreading specimens, 234–236, 238, 241
mounting blocks, 237
Natural History Museum of Los Angeles County, 10–11
night collecting, 218, 224–226
permanent, 233–234, 239–240
photography as tool, 250–254
records and field notes, 232–233
San Diego Museum of Natural History, 10
temporary storage, 230–231

University of California, 9
where to look, 213–218
beetle studies
California, history, xvi, 2–8
California Dung Beetle Project, 129–132, 134
ecological, 58–59
ethical concerns, 218–221
faunistics, 58–59
future of, 23
observing beetles in field, 231, 250, 254
observing captive beetles, 249–250
observing prepared specimens, 238, 240–241
research institutions, 8–11
zoogeographic, 58–59, 94–95
beetles
factors in success of, 52–53
as food, 2–3
species, statistics, xv, 52
Behrens, James H., 12
Berlese funnels, 215, 223–224
Berosus, 148
binoculars, 254
biological controls
for bark beetles, 72, 152
for bees and wasps, 184
blister beetles as, 193
carpet beetles as, 117
for citrus pests, 132–133
dung beetles as, 129
for eucalyptus longhorn beetle, 120–121
for eucalyptus snout beetle, 122
field testing, 134–135
ground beetles as, 72
for Horn Fly, 152
lady beetles as, 94–95, 133–134, 187
leaf beetles as, 94–95, 201–202
for longhorn beetles, 197
research history, xvi
rove beetles as, 72
soldier beetles as, 177
unforeseen consequences, 95
weevils as, 205
bioluminescence, 67, 172, 174–175
bipectinate antennae, 36
birds
as beetle predators, 160, 197
nests as beetle habitat, 71–72, 152, 180, 217–218
Blaisdell, Frank Ellsworth, 18, 19–20
Blaschke, E.L., 6
blister beetle family, 179, **192–194**
antennae, 35
collecting, 213
courtship behavior, 69
defense strategies, 79–80
head, 30
larvae, 50, 64
as parasites, 74
thorax, 38
wings, 40, 42
Blister Beetle, Inflated, 40, 41, 91
blister beetles, false, 179, 194
blistering and burning, as defense strategies, 79–81
blood, circulatory system, 44–45
blue death feigners, 75
Bolboceras obesus, horns, 31
bombardier beetles, defense strategies, 79
Borer(s)
California Flatheaded, 111
Eucalyptus, 118–121, 195, 197
Giant Palm, 182, 183
Hairy Pine, 90, 197
Mesquite, 92
Nautical, 2
Ribbed Pine, 2, 90, 195
borers (wood-boring beetles)
as beneficial, 107
branch or twig, **181–184**
camouflage, 75–76
collecting, 214–215
drag resistance, 29
endosymbiosis, 73
feeding habits, 71
flatheaded, 111, 167
as forest pests, 107
milkweed, 80–81, 195, 202
mouthparts, 35
phoretic species, 74

borers *(continued)*
roundheaded, 195
Bostrichidae, 38, **181–184**
Brachinus
defense strategies, 79
egg-laying, 138
Brachys, feeding habits, 167
Braconidae, 197
brain, morphology, 43
branch-borer family, **181–184**
Brentidae, 207
broom
scale, as beetle habitat, 212, 214
scotch, biological controls for, 205
brushes, camel hair, 227–228
buckthorn, as beetle habitat, 214
buckwheat, as beetle habitat, 214
Buprestid, Golden, 168
Buprestidae, **166–169**, 171
collecting, 213
defense strategies, 77
drag resistance, 29
egg-laying, 63–64
as forest pests, 107, 111
as prey, 184
wings, 40
Buprestis, feeding habits, 167
burrowing beetles, collecting, 210
Burying Beetle, Black, 36, 155
burying beetles, 154–155
collecting, 216
feeding habits, 72
parental care, 69–70
relationships with mites, 74
Byrrhidae, 181

Calico Mountains, 103
California
beetle distribution patterns, 85, 88–94
deserts, 86–87
European colonization, xvi, 2, 3–4
Great Central Valley, 86
islands, 87–88
map, 84
military outposts, xvi, 7–8, 14, 15
mountains, 85–86
native Americans, xvi, 2–3
research institutions, 8–11
Russian occupation, 4–6
California Academy of Sciences, 8–9
California Dung Beetle Project, 129–132, 134
California fauna, 88–89, 93–94
California lilac, as beetle habitat, 213–214
California native plants, as beetle habitat, 213–214
Callidiini tribe, systematics, 58
Calosoma
distribution patterns, 89
egg-laying, 138
feeding captives, 243–244
semilaeve, 140
Calosoma, Common Black, 140
cameras, 250–254
camouflage, as defense strategy, 44, 75–77
Camp Independence, 7, 15
campodeiform larvae, 48, 50
Cantharidae, **176–179**, 186, 194, 199
courtship behavior, 69
feeding habits, 72
cantharidin, defense strategy role, 79–80, 192
Cantharis
consors, 178
fungal infections of, 177
Canthon
impact of exotic species on, 135
morphology, 164–165
reproductive systems, 47
simplex, 128
capitate antennae, 36
captive beetles
caring for, 242–245, 247–249
collecting, 241–242
observing, 249–250
from pet shops, 245–247
Carabidae, **138–141**, 191
antennae, 36
camouflage, 76
collecting, 215
distribution patterns, 89

drag resistance, 29
endangered and threatened species, 105
feeding habits, 71
fossils, 101
mating, 69
mouthparts, 33
sensitive species, 103
cardenolides, defense strategy role, 80–81
carnivores
collecting, 216
feeding habits, 71–72
See also predatory beetles
Carpenteria deposit, 102
Carpet Beetle(s)
Furniture, 116
Varied, 82–83 (photo), 116–117
Carpophilus hemipterus, as pantry pest, 115
carrion beetle family, **154–156**
collecting, 216, 226
feeding habits, 72
fossils, 101
Cascade Range, 90, 91
Casey, Thomas Lincoln, 12, 16–17
Catalogue of the Coleoptera, 13
caterpillar hunters, feeding captives, 243–244
Ceanothus, as beetle habitat, 213–214
Cellucotton, 231
Centrodera, distribution patterns, 90
Cerambycidae, 179, **194–199**, 203
antennae, 36, 37
collecting, 213
defense strategies, 80–81
drag resistance, 29
egg-laying, 64
endangered and threatened species, 106–107
as forest pests, 107
mimicry by, 77–78
mouthparts, 35
native Americans and, 2
oviposition, 47
phoretic species, 74
as prey, 184
systematics, 58
The Cerambycidae of North America, 22–23
Ceratophyus gopherinus, horns, 31–32
Cercocarpus, as beetle habitat, 214
Cerylonidae, 192
chafer family, **162–166**
camouflage, 76
fossils, 101
mouthparts, 35
pupae, 51
Chalcolepidius, 169, 171
Chalcophora, distribution patterns, 90
chamise, as beetle habitat, 214
Chamisso, Adelbert von, 5
Channel Islands, 87–88
Chaudoir, Baron, 14
Chauliognathus, fungal infections of, 177
checkered beetle family, **184–186**, 197, 203
feeding habits, 72
mimicry by, 78
as parasites, 74
Checkered Beetle, Ornate or Common, 185
Chemsak, J.A., 23
Child, J., 14
chitin, 27
Chlaenius, egg-laying, 138
Chlamisus, camouflage, 76–77
chlorocresol crystals, 231
Chrysochus colbaltinus, 81, 202
Chrysolina
hyperici, 201–202
quadrigemina, 201–202
Chrysomelidae, 189, 199, **200–203**
camouflage, 44, 76
defense strategies, 81
as forest pests, 121–122
larvae, 48
mouthparts, 33
Chrysomelinae, 189
Chrysophana, feeding habits, 167
Chrysothamnus, as beetle habitat, 212, 214

Cicindela
californica, 140
camouflage, 76
feeding captives, 244
latesignata obliviosa, 103
mandibles, 34
mating, 69
ohlone, 105
oregona, 36
tranquebarica, 103
Cicindelidae, 139
Cigarette Beetle, as household pest, 113
circulatory system, morphology, 44–45
citrus pests, biological controls for, 132–133
classes, defined, 56
classification
Linnaean system, 53–54
systematics, 57–59
taxonomy, 56–57
Classification of the Coleoptera of North America, 14
clavate antennae, 36
Cleridae, **184–186**, 197, 203
feeding habits, 72
introduction to California, 4
as pantry pests, 114–115
Click Beetle, Eyed, 77, 170, 197
click beetle family, **169–171**, 197, 199
collecting, 214
distribution patterns, 89, 91
thorax, 38
wings, 40
wireworms, 50, 169–170
click beetles, false, 169, 171
climate, role in beetle collecting, 210–214
clown beetle family, **151–153**
collecting, 216
feeding habits, 71–72
clown beetles, false, 153
clypeus, function, 34–35
Coachella Valley, 94
Coast Ranges, 85, 90, 93, 211
Coccinellidae, **186–189**
as biological control agents, 94–95, 133–134, 187
feeding habits, 72
larvae, 49
native Americans and, 3
Coelus, distribution patterns, 93–94
Coenonycha
dimorpha, 55
distribution patterns, 93
Coleoptera order, 56, 57
collection boxes, 239–240
Colorado Desert, 86–87, 92, 94, 210, 212
Colydiidae, 184
comb-clawed beetles, 141
commensalism, defined, 73
common names, 55
Commonwealth Scientific and Industrial Organization of Australia, 129
Coniontis remnans, fossils, 101
Copris pristinus, fossils, 101
copulatory organs
function, 68–69
identification role, 42, 240–241
morphology, 42, 46–47
Coquillet, D., 133
coronnes, 3
Corylophidae, 189
Cotinis
life history, 163
mutabilis, 40, 44, 243, 244
Cottony Cushion Scale, biological controls for, 133
courtship behavior. *See* mating
coxae, morphology, 27, 38–39
Cremastocheilus
exoskeleton, 28–29
feeding habits, 72
life history, 163
Cryptochetum iceryae, as biological control agent, 133
Cucujidae, 197
Cucujus clavipes puniceus, 29–30
Curculionidae, 184, **204–207**
antennae, 36
collecting, 216
distribution patterns, 91
economic importance, 107, 109–111, 204–205

egg-laying, 63–64
as eucalyptus pests, 122
larvae, 48
mouthparts, 35
as pantry pests, 115–116
parental care, 70
stridulation, 68
wings, 41
cuticle, function, 28
Cybister, legs, 39
Cyclocephala, life history, 163
Cyclophala, as turf pests, 164
Cymatodera, feeding habits, 184
Cypriacis aurulenta, as forest pest, 168
Cysteodemus armatus
distribution patterns, 91
wings, 40

darkling beetle family, 141, **189–192**, 203
collecting, 214, 217
defense strategies, 79, 249–250
distribution patterns, 91, 92, 93–94
feeding captives, 243, 248–249
fossils, 101
mealworms, 50, 190, 243, 245–249
mimicry of, 77
native Americans and, 3
wings, 40, 42
darkling beetles, false, 171, 192
Dascillidae, 171
Dastinae, 186
death watch beetles, 181, 184
courtship behavior, 68
as household pests, 113–114
decomposition
role of beetles, 70, 107
See also dung beetle family
defense strategies
blistering and burning, 79–81
bold designs and colors, 77, 79
camouflage, 44, 75–77
chemical, 79–81
mimicry, 77–78
observing, 249–250
thanatosis, 75, 76
Dejean, Pierre Francois Marie Auguste, 6
DeLorme Mapping Company, 232
Dendroctonus, 204
brevicomis, 41, 111
as forest pests, 109–111
jeffreyi, 111, 112
ponderosae, 111, 112
pseudotsugae, 111
valens, 110, 111
Dermestes, 181
introduction to California, 4
marmoratus, 180
Dermestid, Common Carrion, 180
Dermestidae, **179–181**
feeding habits, 72
fossils, 103
as household pests, 116–117
introduction to California, 4
Derobrachus geminatus geminatus, distribution patterns, 92
Desert Ironclad Beetle, exoskeleton, 28
desert ironclad beetles
observing, 250
thanatosis, 75
deserts
adaptations to, 40–41, 91
beetle distribution patterns, 86, 91–92
California, 86–87
collecting beetles in, 210, 211, 212
Desmocerus
californicus californicus, 106–107
californicus dimorphus, 106–107
deuterocerebrum, function, 43
Dichelonyx
camouflage, 76
as horticultural pests, 164
digestive system
endosymbiosis, 73
morphology, 43–44
Dinapate wrightii, 182, 183
dip nets, 222
Diphyllostomatidae, 165
Diplotaxis
as horticultural pests, 164
as nocturnal, 163
diseases of insects, 177
Disonycha, 203

distribution patterns
about, 88–89
California fauna, 88–89, 93–94
in deserts, 86
on islands, 88
southern Sonoran fauna, 88–89, 91–92
systematic tools, 58–59
Vancouveran fauna, 88–91
Diving Beetle(s)
Mono Lake Hygrotus, 103
Sunburst, 146
Yellow-spotted, 146
Douglas-fir Beetle, 111
drag resistance, adaptations to, 29–30
Dried Fruit Beetle, as pantry pest, 115
drought, adaptations to, 40–41, 91
Drugstore Beetle, as pantry pest, 113–114
Dryopidae
collecting, 215–216
respiratory systems, 46
dung beetle family, 151, **162–166**
antennae, 37
California Dung Beetle Project, 129–132, 134
collecting, 216–217, 226
feeding habits, 71
fossils, 101
horn size, 33
importance of, 128–129
mouthparts, 33
parental care, 69–70
reproductive systems, 47
Dung Beetle, Spotted, 153, 217
Dunn, George W., 12
Dynastes hercules, 162
Dytiscus marginicollis, fossils, 101
Dytiscidae, **144–147**, 151
fossils, 101
hydrodynamic adaptations, 29
Dytiscus
distribution patterns, 89
legs, 39

earth-boring beetles, thorax, 38
echolocation, role of antennae, 35–36
ecological studies, 58–59
ectoparasites, 74
ectysial suture, 47
Edrotes, distribution patterns, 92
egg-laying. *See* oviposition
Elacatis, 186
Elaphrus viridis, distribution patterns, 105
Elater, distribution patterns, 89
Elateridae, **169–171**, 197, 199
antennae, 36
collecting, 214
defense strategies, 77
larvae, 50
thorax, 38
elateriform larvae, 48, 50
Eleodes, 189, 191
defense strategies, 79
distribution patterns, 91, 92
feeding captives, 243
Elephant Beetle(s), 162
Sleeper's, 32, 92
Ellychinia, 175
californica, 175
Elmidae, 151, 216
elytra
morphology, 40–42
role in success of beetles, 52
taxonomic role, 56
elytral suture, morphology, 40
endangered and threatened species, 98, 104–107
See also sensitive species
Endangered Species Act (ESA), 104
Endomychidae, 189, 203
endosymbiotic microorganisms, 73
engraver beetles, 109–111
Engraver(s)
Fir, 110
Pine, 111, 112
Enoclerus, feeding habits, 184
Ergates
spiculatus, 2, 74, 90, 197, 229
spiculatus spiculatus, 34
distribution patterns, 89, 90
Eriogonum fasciculatum, as beetle habitat, 214
Erotylidae, 189, 192, 203
eruciform larvae, 48, 49
Eryniopsis, as beetle pathogen, 177

Eschscholtz, Johann Freidrich, 5–6, 14
Essig Museum of Entomology, 9
ethyl acetate, 229–230
eucalyptus, beetle pests of, 117–122
Eucinetidae, 181
Eucnemidae, 169, 171
Euoniticellus intermedius, California
Dung Beetle Project, 132
Eupagoderes, distribution patterns, 92
Euphoria, life history, 163
Eusattus, distribution patterns, 91
Euthysanius, 171
lautus, 36
evolution of beetles
fossils, 88, 99–103
role of plants, 52–53
exarate pupae, 51
Exema, camouflage, 76–77
exoskeleton
molting, 47, 64
morphology, 26–29
exotic beetle species
Asian Longhorn Beetle, 126–128
from Australia, 117–122, 129, 133–134
beneficial, 98
dung beetles, 129–132, 134, 216–217
impact of, xvi, 94–95, 98, 134–135
introduced by European settlers, 3–4, 98
Japanese Beetle, 123–126
See also biological controls
eyes
larval, 47–48
morphology, 31

Face Fly, biological controls for, 129
Fall, Henry Clinton, 15, 18–19
families, defined, 56
Farrallon Islands, 87
faunistic studies, 58–59
FBI (fungi, bacteria, insects), 70
feeding habits
of captive mealworms, 247–249
of carnivores, 71–72, 216
of herbivores, 70–71, 213–214, 243
myrmecophiles, 73
femurs, morphology, 27, 39
filiform antennae, 36
fir, as beetle habitat, 214
firefly family, 30, **174–176**, 178
Fischer, F., 6
flabellate antennae, 36
flagella, morphology, 35, 36
flags (branch tips), 64, 197
flea beetles, 203
flies
as biological control agents, 133–134
biological controls for, 72, 129
tachinid, 197
Flour Beetle, Confused, 115
flower beetles
antlike, 38, 80, 186, 192, 203
collecting, 210, 211, 213–214
minute, 151
shining, 188
soft-winged, 40, 186, 203
tumbling, 213, 214
fly maggots, predatory, 160
forceps, 227–228
foregut, function, 44
forest pests
about, 107
Australian, 117–122
feeding habits, 71
research, 21
See also specific forest pests
Fort Crook, 15
Fort Ross, 4, 6
Fort Sutter, 6
Fort Tejon, 7, 15
Fort Yuma, 15
fossil beetles
about, 99
in asphalt deposits, 99–102
from California islands, 88
coevolution with plants, 52–53
in mineral deposits, 103
spider beetles, 102–103
frass, 44
Frazier Mountain, 103
Fruit Beetle, Green, 40
Fuchs, Charles, 12
fungal feeders, 70, 107–108

fungal pathogens, 177
fungi, as beetle habitat, 215
fungus beetles
  handsome, 189, 203
  minute, 189
  pleasing, 189, 192, 203
  round, 188
  shining, 153

ganglia, morphology, 43
genera, defined, 54, 56
geniculate antennae, 36
Geotrupidae, 166
gills, physical, 46
gin-traps, 51, 246
girdlers, egg-laying, 64
Glaphyridae, 78, 166
Glaresidae, 165
glowworm family, **174–176**, 178
  bioluminescence, 67
  head morphology, 30
  wings, 40
  See also phengodid beetle family
Glowworm(s)
  California, 175
  California Banded, 68
  Pink, 24–25 (photo), 65, 66
  Western Banded, 173 (photo), 208–209 (photo)
Goliathus, 162
Gonipterus scutellatus, as eucalyptus pest, 122
gorse, biological controls for, 205
Grain Beetle, Sawtoothed, 115
grasshoppers, biological controls for, 74, 184, 193
Great Basin, 86–87
Great Basin fauna, 92, 212
Great Central Valley
  about, 86
  endangered and threatened species, 105, 106–107
  fauna, 92
Green Fig Beetle
  excrement use by, 44
  feeding captives, 243, 244
Ground Beetle, Delta Green, 105
ground beetle family, **138–141**, 191
  antennae, 35
  collecting, 215, 217
  defense strategies, 79
  distribution patterns, 91
  drag resistance, 29
  endangered and threatened species, 105
  feeding captives, 243
  feeding habits, 71, 72
  fossils, 101
  larvae, 50
  legs, 39
  mouthparts, 33, 35
grubs
  feeding captives, 194
  as food, 2–3
  ladybug, 49
gula, defined, 35
Gypsy Moth, biological controls for, 117
Gyrinidae, 29, **141–144**, 147
Gyrinus plicifer, 143

Haematobia irritans, 152
  biological controls for, 129
Haldeman, S.S., 13
Haliplidae, 39, 147
Ham Beetle, Redlegged, 114–115
ham beetles
  collecting, 216
  feeding habits, 72
  introduction to California, 4
hand lenses, 231
Harford, W.G.W., 12
Harvard University, Museum of Comparative Zoology, 14
head
  horns, 31–33
  larval, 47–48
  morphology, 27, 30
heart, morphology, 44–45
herbivores
  collecting, 213–214
  feeding captives, 243
  feeding habits, 70–71
Hercules Beetle, 162
hide beetles, 166
  collecting, 217
  feeding habits, 72

high-temperature adaptations, 40–41, 91
hindgut, function, 44
Hippodamia convergens, 96–97 (photo), 188
as biological control agents, 134
larvae, 49
life cycle, 63
Hister, 152
Histeridae, **151–153**
collecting, 216
feeding habits, 71–72, 73
wings, 40
holotypes, defined, 55
Hoplia, as horticultural pests, 164
hormones
pheromones, 37, 68, 128
role in horn size, 33
Horn, George Henry, 7, 14–16
Horn Fly, biological controls for, 129, 152
horns, head
morphology, 31–32
size, 32–33
horticultural pests
from Australia, 117–122
feeding habits of, 70–71
introduction to California, 94–95
See also specific horticultural pests
household pests, 113–117, 218
See also specific household pests
humahnana, 3
Hybosoridae, 166
Hybosorus illigeri, distribution patterns, 94
Hydraenidae, 151
hydrodynamic adaptations, 29
Hydrophilidae, **147–151**, 153
collecting, 217
feeding captives, 245
fossils, 101
oviposition, 47
Hydrophilus triangularis, 47, 101, 150
Hydroscaphidae, 158
Hygrotus artus, 103
Hypera, 205
hypermetamorphosis, 64, 193, 194
hypognathous mouthparts, defined, 35
Icerya purchasi, biological controls for, 133
Ichneumonidae, 197
identifying beetles, 240–241
Imperial Academy of Sciences (Russia), 6
Insecta class, 56, 57
inspection programs, for Japanese Beetle, 125–126
instars, defined, 64
International Code of Zoological Nomenclature, 55
Ipochus
distribution patterns, 93
fasciatus, 93
Ips
emarginatus, 112
paraconfusus, 112
as pests, 111–112, 204
pini, 111, 112
Ips(s)
California Fivespined, 112
Emarginate, 112
islands, California, 87–88

Japanese Beetle, **123–126**
Jepson Prairie Reserve, 105
jewel beetle family, **166–169**
Johnston's organ, 35–36
June beetle family, **162–166**
endangered and threatened species, 105–106
phoretic species, 74
stridulation, 68
June Beetle(s)
Dusty, 36, 51
Mount Hermon, 105–106
Striped, 3
Ten-lined, 1 (photo), 55, 165

killing jars and agents, 229–230
klamathweed, biological controls for, 201–202
knapweed, biological controls for, 205
Koebele, Albert, 12, 133

La Brea Tar Pits, 99–102
labeling specimens, 237

labia, functions, 34
lady beetle family, **186–189**
as biological control agents, 72, 94–95, 133–134
camouflage, 76
larvae, 49
life cycle, 63
native Americans and, 3
pupae, 51
Ladybird, Convergent, 96–97 (photo), 134, 188
ladybird beetles or ladybugs. *See* lady beetle family
Lake Cahuilla, 93
lamellate antennae, 36
Lampyridae, **174–176**, 178
antennae, 36
head morphology, 30
Languriidae, 169, 171
larvae
coarctate, 194
as food, 2–3
grubs, 2–3, 49, 194
life cycles, 64–65
mealworms, 50, 190, 243, 245–249
morphology, 47–50
preserving specimens, 238–239
wireworms, 50, 169–170
Lasioderma serricorne, 113
leaf beetle family, 189, 199, **200–203**
antlike leaf beetles, 186
as biological control agents, 94–95, 201–202
biological controls for, 72
camouflage, 76–77
feeding habits, 70–71
larvae, 49
legs, 39
mouthparts, 33, 35
pupae, 51
warty leaf beetles, 201
leaf miners and grazers, 63, 200–201, 204–205
leafcutter bees, 196
Leatherwing Beetle, Brown, 178
LeConte, John L., 7, 11–14
LeConte, Major John E., 12
Leech, Hugh Bosdin, 21–22
legs
larval, 48
morphology, 38–39
Leiodidae, 74, 188
Lepidospartum squamatum, as beetle habitat, 212, 214
Liatongus
californicus californicus, 128
impact of exotic species on, 135
morphology, 164–165
lichens, as beetle habitat, 215
Lichnanthe, mimicry by, 78
lightning bugs. *See* firefly family; glowworm family
lights, beetles attracted to, 218, 224–226, 227
Limonius, distribution patterns, 89
Linnaeus, Carolus, 53–54
Linsley, Earle Gorton, 22–23
Lion Beetle, 77–78, 195
livestock pests
biological controls for, 129, 152
blister beetles, 192
lizard beetles, 169, 171
longhorn beetle family, 179, **194–199**, 203
antennae, 37
collecting, 213, 214–215
defense strategies, 75
distribution patterns, 89–90, 91–92, 93
drag resistance, 29
egg-laying, 64
endangered and threatened species, 106–107
eucalyptus, 118–121
as food, 2
as forest pests, 107
legs, 39
mimicry by, 77
mouthparts, 35
as prey, 184
research, 21
stridulation, 68
systematics, 58
wedge-shaped beetles and, 74
wings, 40

Longhorn Beetle(s)
Asian, **126–128**
Cactus, 47, 77, 92
California Elderberry, 106–107
Valley Elderberry, 106–107
longhorn beetles, false, 199
Lorquin Entomological Society, 11
Lucanidae, 165, 184
endosymbiosis, 73
larvae, 49
luciferin enzyme, 67
lumber pests. *See* forest pests
lupine *(Lupinus)*, as beetle habitat, 214
Lycidae, 35, 176, 178
Lyctidae. *See* Bostrichidae
Lymanatria dispar, biological controls for, 117
Lytta
magister, 193
vesicatoria, 192

Malpighian tubules, function, 44
mammals
as beetle predators, 160
nests as beetle habitat, 71–72, 102–103, 152, 180, 211, 217–218
mandibles
functions, 34
morphology, 27
Mannerheim, Carl Gustov von, 6, 14
marsh beetles, 188
mating
abdomen adaptations, 42
antennae role, 35
bioluminescence role, 67
courtship behaviors, 68–69
horn size role, 32–33
pheromone role, 37, 68
stridulation role, 41, 68
See also reproductive systems
maxilla, functions, 34
May beetle family, **162–166**
mealworms
morphology, 50, 190
from pet shops, 50, 243, 245–247
raising, 247–249
Mealworm(s), Common or Yellow, 245–248
Mealworm(s), Super, 248–249
mealy bugs, biological controls for, 133
Mediterranean climate, 210
Megachilidae, 196
Megasoma
distribution patterns, 92
elephas, 162
sleeperi, 32, 92
Melandryidae, 171, 192
Melanophila californica, 111
mellow bugs. *See* whirligig beetle family
Meloidae, 179, **192–194**
courtship behavior, 69
head morphology, 30
wings, 40
Melolontha decemlineata, nomenclature, 55
Melsheimer, F.E., 13
Melyridae, 40, 186, 203
Ménétriés, E., 6
mentum, defined, 35
mesothorax, 37–38
mesquite, as beetle habitat, 214
metallic wood-boring beetle family, **166–169**, 171
collecting, 213, 214–215
defense strategies, 77
distribution patterns, 90
drag resistance, 29
egg-laying, 63–64
as forest pests, 107
as prey, 184
wings, 40
metamorphosis
complete, 47–50, 62–64
hypermetamorphosis, 64, 193, 194
paedogenesis, 65
metathorax, 37–38
Micromalthidae, 186
Microphotus angustus, paedogenesis, 24–25 (photo), 66
middens, rodent, 102–103
nests as beetle habitat, 152, 211, 217–218
midgut, function, 44

military outposts, entomological studies at, xvi, 7–8, 14, 15
Milkweed Beetle, Blue, 81, 202
mimicry, as defense strategy, 77–78, 195
Miocryptorhopalum kirkbyae, fossils, 103
Mission San Diego, 3
missions, Spanish, xvi, 2, 3–4
mites, relationships with beetles, 74
Mojave Desert, 86–87, 92, 210, 212
molting, 47, 64
Monarthrum scutellare, 108
Moneilema semipunctata
 distribution patterns, 92
 mimicry by, 77
 oviposition, 47
moniliform antennae, 36
Monochamus
 maculosus, 2
 scutellatus, 3
 scutellatus oregonensis, 37, 195, 197
Monoxia, 200
Mordellidae, collecting, 213
morphology
 abdomen, 42
 antennae, 35–37
 circulatory system, 44–45
 digestive system, 43–44
 drag resistance, 29–30
 exoskeleton, 27–29
 eyes, 31
 head, 30
 horns, head, 31–33
 larvae, 47–50, 64
 legs, 38–39
 mouthparts, 33–35
 nervous system, 43
 pupae, 50–51
 reproductive system, 42, 46–47, 68–69
 respiratory system, 45–46
 role in beetle success, 52
 role in classification, 54
 thorax, 37–38
 wings, 40–42
moss beetles, minute, 151
mosses, as beetle habitat, 215
Motschulsky, V.I., 6, 14
mountain mahogany, as beetle habitat, 214
mountains, California, 85–86
mounting beetle specimens, 234–238, 241
mounting blocks, 237
mouthparts
 larval, 48
 morphology, 33–35
Musca autumnalis, biological controls for, 129
mutualism, defined, 72–73
mycangia, 70, 107–108
myrmecophiles, 73

Nagano tiger beetle collection, 11
National Museum of Natural History, 16
Native Americans, role of beetles, xvi, 2–3
natural enemies. *See* biological controls
Natural History Museum of Los Angeles County, 10–11, 87
Necrobia
 collecting, 216
 introduction to California, 4
 as pantry pests, 114–115, 184
 rufipes, 114–115
Nemonychidae, 207
nervous system, morphology, 43
nets, collecting, 221–222
net-winged beetles, 35, 176, 178
New California, 5
New Helvetia, 6
Nicrophorus
 Nigrita, 36, 155
 parental care, 69–70, 154–155
Nitidulidae
 collecting, 213
 as pantry pests, 115
 wings, 40
nomenclature, binomial, 53–55
Nosodendridae, 181
Noteridae, 147
Notoxus, 80, 192
Nyctoporis carinata, antennae, 36

oak, as beetle habitat, 214
ocelli, morphology, 47–48
Ochodaeidae, 166
Octinodes, 171
Oedemeridae, 179, 194
Onitis alexis, California Dung Beetle Project, 131
Onthophagus, 164–165
  everestae, 101
  gazella, 130–131
  reproductive systems, 47
  taurus, 33, 130–131
Ophraella, pupation, 200
Ophryastes, distribution patterns, 91, 92
orders, defined, 56
Oryzaephilus surinamensis, 115
Ostoma pippingskoeldi, 36
Othniinae, 141
ovaries, morphology, 47
oviducts, morphology, 47
oviposition, 42, 47, 62–64, 69
ovipositors, morphology, 42
oxygen transport system, 45–46

Pacific Coast Entomological Society, 9
Pacific Northwest mountain region, 85
paedogenesis, 65–66
palps, labial and maxillary
  functions, 34
  morphology, 27
pantry pests
  collecting, 218
  controlling, 116
  important species, 113–116
  See also specific pantry pests
Paracotalpa, distribution patterns, 93
Parasite Beetle, Beaver, 74
parasitic beetles
  about, 74
  larvae, 50
  metamorphosis, 64
parasitism, defined, 73
Parathyce
  life history, 163
  palpalis, 36
parental care, 69–70, 154–155
parthenogenesis, defined, 69
Peach Beetle, feeding captives, 243
Peale, Titian, 14
pedicels, morphology, 35, 36
Peninsular Ranges, 85, 90, 91, 211
pet shop beetles, 50, 243, 245–247
Phalacridae, 151, 188
Phanaeus labreae, fossils, 101
Phengodid, California, 36
Phengodidae (phengodid beetle family), **172–173**, 176, 179
  antennae, 36, 37
  bioluminescence, 67
  feeding habits, 72
  paedogenesis, 65
  pheromones, 68
  wings, 40
pheromones
  detection by antennae, 37, 68
  role in mating, 68
  as trapping aid, 128
Phloeodes, distribution patterns, 93
Phoracantha
  as forest pests, 195, 197
  recurva, 118–121
  semipunctata, 118–121
phoresy, defined, 73–74
photographing beetles, 250–254
Phrixothrix, 172
phyla, defined, 56
Phyllophaga, life history, 163–164
Phymatodes vulneratus, 21
Pickering, Dr. Charles, 14
pill beetles, 181
pine *(Pinus),* as beetle habitat, 214
Pine Beetle(s)
  Jeffrey, 111, 112
  Mountain, 111
  Western, 41
  Western Banded, 111
pins, mounting, 234–236
plant beetles, soft-bodied, 171
plant diseases, beetles as vectors, 201
plants
  as beetle habitat, 213–214
  coevolution with beetles, 52–53
plastron breathing, 46
plate-thigh beetles, 181

Platypsyllus castoris, as ectoparasites, 74
Pleocoma, **158–162**
australis, 161
badia badia, 136–137 (photo), 161
conjugens conjugens, 106
distribution patterns, 90
Pleocomidae, **158–162**, 166
antennae, 37
endangered and threatened species, 106
larvae, 64
Pleotomus nigripennis, antennae, 36
pleuron, morphology, 42
plumose antennae, 36
Podabrus, fungal infections of, 177
pollen-feeding beetles, collecting, 210, 211, 213–214
Polycaon Beetle, Black, 65
Polycaon stouti, metamorphosis, 65
Polyphylla
Barbata, 105–106
Crinita, 3
Decemlineata, 1 (photo), 55, 165
as horticultural pests, 164
life history, 163
phoretic species, 74
Popillia japonica, **123–126**
powder-post beetle family, **181–184**
predaceous diving beetle family, **144–147**, 151
collecting, 215–216
distribution patterns, 89, 91
fossils, 101
hydrodynamic adaptations, 29
legs, 39
respiratory systems, 46
predatory beetles
drag resistance, 29
feeding captives, 243
feeding habits, 71–72
larvae, 48, 50
mouthparts, 35
Prionus, California *(Prionus californicus),* 198 (photo)
antennae, 36
economic importance, 107, 197
as food, 2
phoresy, 74
studies of, 17
prognathous mouthparts, defined, 35
pronotum, 27, 38
propylene glycol, 227
Prosopis, as beetle habitat, 214
prothetely, 170
prothorax, 37–38
protocerebrum, function, 43
proventriculus, function, 43–44
Psephenidae, collecting, 216
Pseudococcidae, biological controls for, 133
pseudoscorpions, 73–74
Pterostichus, egg-laying, 138
Ptinus
Clavipes, 114
introduction to California, 4
priminidi, 102–103
pupae
life cycles, 65
morphology, 50–51
preserving specimens, 238–239
purple loosestrife, biological controls for, 205
push ups, defined, 211

Quercus, as beetle habitat, 214

rabbit brush, as beetle habitat, 212, 214
railroads, entomological surveys for, 7, 8
rain beetle family, 136–137 (photo), **158–162**, 166
antennae, 37
distribution patterns, 90
larvae, 64
research, 23
Rain Beetle(s)
Santa Cruz, 106
Southern, 161
Rancho La Brea, 99–102
reflex bleeding, 80
reproductive systems
female, 42, 47, 69
male, 42, 46–47, 68–69, 240–241
respiratory systems
adaptations to water, 46

morphology, 45
Rhagium inquisitor
distribution patterns, 90
feeding habits, 195
as food, 2
Rhipiceridae, antennae, 36
riffle beetles, 151
collecting, 216
respiratory systems, 46
Riley, C.V., 133
Ripiphoridae (ripiphorid beetle family)
as parasites, 74
wings, 40
Rivers, James, 12
robber flies, 160
rodent nest dwellers, 152
Rodolia cardinalis, as biological control agent, 133–134
Root Borer, St. John's, 201
Rosalia, distribution patterns, 89
rove beetle family, **156–158**
abdomen, 42
antennae, 35
collecting, 215, 216, 217
distribution patterns, 91
feeding habits, 71, 72
fossils, 101
pupae, 51
wings, 40
Rove Beetle, Pictured, 157
Russian occupation, Northern California, xvi, 2, 4–6

Sacramento Valley, 86
Salpingidae, 141, 186
San Diego Museum of Natural History, 10
San Gabriel mission, 132
San Joaquin Valley, 86, 92
San Pedro Martir, 91
sand dune dwellers
collecting, 212–213, 216
distribution patterns, 93–94
endangered and threatened species, 105–106
sensitive species, 104
Sandalus, antennae, 36
Santa Cruz Island, 87
Santa Rosa Mountains, 93
sap beetles
collecting, 213
wings, 40
Saprinus lugens, 153
Sawyer(s)
Black Pine, 3
Oregon Fir, 37, 195, 197
Pine, xvi, 34, 74, 90, 197, 229
Spotted Pine, 2
scapes, morphology, 35, 36
Scaphidiinae, 153
Scaphinotus, distribution patterns, 90
Scarab, Sacred, 162–163
Scarabaeidae (scarab beetle family), 151, **162–166**
antennae, 36
ant-loving scarabs, 28–29
bumblebee scarabs, 166
collecting, 213, 216
distribution patterns, 93
dung decomposing role, 128–132
earth-boring scarabs, 31, 166
endangered and threatened species, 105–106
enigmatic scarabs, 165
Evans collection, 11
feeding habits, 70–71, 72, 73
as food, 3
fossils, 101
horn size, 33
larvae, 48, 49
legs, 39
mimicry by, 78
mouthparts, 33, 35
parental care, 69–70
phoretic species, 74
sand-loving scarabs, 166
scavenger scarabs, 94, 166
stridulation, 68
wings, 40
scarabaeiform larvae, 48, 49, 50
Scarabaeus sacer, 162–163
scavenger beetles
collecting, 227
feeding habits, 71
scientific names. *See* nomenclature, binomial

Scirtidae, 188
sclerites, 27
sclerotin, 27
Scolytus ventralis, 110
scutellum, morphology, 27, 40
Scyphophorus
distribution patterns, 91
yuccae, 36, 206
sensitive species, 103–104
See also endangered and threatened species
Serica
as horticultural pests, 164
kanakoffi, 101
life history, 163
serrate antennae, 36
setae, 28
Sierra Foothill Range Field Station, 130
Sierra Nevada mountain region, 85, 90, 91, 210, 215
Silphidae, **154–156**
antennae, 36
feeding habits, 72
fossils, 101
parental care, 69–70
relationships with mites, 74
Silvanidae, as pantry pests, 115
Sinodendron
distribution patterns, 90
rugosum, 31, 32, 50, 60–61 (photo), 89
Sitophilus
granarius, 3, 115–116, 204
oryzae, 3, 115, 204
skiff beetles, 158
skin beetle family, **179–181**
collecting, 216
feeding habits, 72
fossils, 103
as household pests, 116–117
introduction to California, 4
snail eaters, distribution patterns, 90
Snout Beetle, Eucalyptus, 122
snout beetle family, **204–207**
pine-flower snout beetles, 207
soldier beetle family, **176–179**, 186, 194, 199
courtship behavior, 69
feeding habits, 72
Sonoran fauna, southern, 88–89, 91–92
Spanish Fly, 192
species, defined, 54, 56
spermatheca, morphology, 47, 69
Sphaeridium scarabaeoides, collecting, 217
Sphaeritidae (sphaeritid beetle family), 153, 156
Spider Beetle, Brown, 114
spider beetle family
fossils, 102–103
introduction to California, 4
as pantry pests, 114
spiders, as beetle predators, 197
spiracles
functions, 46
morphology, 45
Spondylis, distribution patterns, 90
St. John's wort, biological controls for, 201–202
Stag Beetle, Rugose, 60–61 (photo), 184
distribution patterns, 89
horns, 31, 32
larvae, 50
stag beetles, 165
distribution patterns, 90
endosymbiosis, 73
larvae, 49
stag beetles, false, 165
Stanford, Governor Leland, 15
Staphylinidae, 153, **156–158**
antennae, 35
collecting, 215
feeding habits, 71, 73
fossils, 101
larvae, 48
starthistle, biological controls for, 205
Stegobium paniceum, 113–114
stem and leaf miners, 167
Stenomorpha, distribution patterns, 91
Stenotrachelidae, 199
sterna, morphology, 42
stink beetles, 189, 191
stridulation, 41, 68
subelytral space, 40–41, 52

subspecies, defined, 54
sunflower family, as beetle habitat, 213–214
Surprise Valley, 15
Sutter, John, 6
sutures, 27
sweep nets, 222
symbiosis, defined, 72
symbiotic relationships, 72–74
systematics, 57–59

taenidia, morphology, 45
Taphrocerus, feeding habits, 167
tarsi, morphology, 27, 39
taxa, defined, 56
taxonomy, 56–57
Tehachapi Mountains, 103
telephone-pole beetles, 186
Tenebrio
  mealworms, 190
  molitor, 50, 243, 245–249
Tenebrionidae, 141, **189–192**, 203
  antennae, 36
  collecting, 214
  distribution patterns, 91
  fossils, 101
  larvae, 48
  mimicry of, 77
  native Americans and, 3
  as pantry pests, 115
  thanatosis, 75
  wings, 40
teneral, defined, 66
tepis, 3
terga, morphology, 42
testes, morphology, 46–47
Tetracha carolina carolina, distribution patterns, 92
Tetraopes, 80–81, 195
thanatosis, 75, 76
Thermonectus marmoratus, 146
Thinopinus pictus, 157
thistles, biological controls for, 205
thorax
  larval, 48
  morphology, 27, 37–38
threatened species. *See* endangered and threatened species
throscid beetles, 171
tibias, morphology, 27, 39
tiger beetle family, **138–141**
  camouflage, 76
  collections, 11, 215, 221–222
  defense strategies, 77
  distribution patterns, 92
  endangered and threatened species, 105
  feeding captives, 244
  feeding habits, 71
  legs, 39
  mandibles, 34
  mating, 69
  mouthparts, 33
Tiger Beetle(s)
  Oblivious, 103
  Ohlone, 105
  Oregon, 36
  San Joaquin Valley, 103
tiger beetles, false, 141
Tortoise Beetle, Eucalyptus, 121–122
tortoise beetles, 44, 76, 201
tracheae, morphology, 45
Trachymela sloanei, 121–122
Tragosoma desparius
  biological controls for, 197
  distribution patterns, 90
Transverse Ranges, 85, 90, 91, 211
traps
  for Asian Longhorn Beetles, 128
  bait, 128, 226
  flight intercept, 227, 228
  lights, 218, 224–226, 227
  pan, 227
  pitfall, 226–227
tribes, defined, 56
Tribolium confusum, as pantry pest, 115
Trichodes
  larvae, 184
  ornatus, 185
Trigonoscuta, distribution patterns, 93
tritocerebral lobes, function, 43
triungulins, 50, 64, 194
trochanters, morphology, 27
trochantin, morphology, 38
Trogidae, 166
  collecting, 217

Trogidae *(continued)*
feeding habits, 72
Trogossitidae
antennae, 36
feeding habits, 72
trout-stream beetles, 191
Trox gemmulatus, collecting, 217
Tschernikh, G., 6
turf pests, 164
Turpentine Beetle, Red, 110, 111
twig-borer family, 38, **181–184**

Ulochaetes
distribution patterns, 90
leoninus, 77–78, 195
United States Department of Agriculture, 129
University of California
agricultural research, 9
California Dung Beetle Project, 129–132, 134
field stations, 87, 130
urogomphi, defined, 49

Van Dyke, Edwin Cooper, 18, 20–21, 22
Vancouveran fauna, 88–91
vasa differentia, morphology, 46–47
Vedalia Beetle, as biological control agent, 133–134
vermiform larvae, 48, 50
vernal pools, 105
Vosnesensky, Ilya Gavrilovich, 6

Walker Basin, 92
Warner's Ranch, 15
water beetles
abdomen, 42
burrowing, 147
collecting, 211, 215–216, 222, 242
crawling, 39, 147
elytra adaptations, 41
feeding captives, 244
See also predaceous diving beetle family; whirligig beetle family
Water Beetle(s)
Giant Black, 101, 150
Giant Green, 101
water beetles
hydrodynamic adaptations, 29
long-toed, 46, 215–216
respiratory system adaptations, 46
thorax, 38
water pennies, collecting, 216
water scavenger beetle family, **147–151**
antennae, 35
collecting, 215–216
feeding captives, 243
fossils, 101
hydrodynamic adaptations, 29
legs, 39
respiratory systems, 46
Water Scavenger, Giant, 47, 48
wedge-shaped beetles, as parasites, 74
weevil family, **204–207**
alfalfa weevils, 205
as biological control agents, 205
collecting, 216
distribution patterns, 91, 92, 93
egg-laying, 63–64
feeding habits, 70–71
as food, 3
fungus weevils, 207
larvae, 50
mouthparts, 35
straight-snouted weevils, 207
toothed snout weevils, 207
wings, 42
Weevil(s)
Boll, 94
Grain, 3, 115–116, 204
Rice, 3, 115, 204
Yucca, 36, 206
whaling ships, introduction of exotic beetles, 4
whirligig beetle family, **141–144**, 147
abdomen, 42
antennae, 35–36
biological controls for, 72
collecting, 215
hydrodynamic adaptations, 29
legs, 39
mouthparts, 35
Wilkes, Captain Charles, 14

wings
absence of, 91
morphology, 27, 40–42
wireworms, 50, 169–170
See also click beetle family
wireworms, false, 190
Wittick, J., 14
wood (dead and stumps), as beetle habitat, 214–215, 242
wood-boring beetles. *See* borers
wounded-tree beetles, 181
Wrangell, F.P., 6

Xantus de Vesey, John, 7
Xylotrechus nauticus, as food, 2

Zarhipis integripennis, 36, 68, 173 (photo), 208–209 (photo)
Zayante sand hills, 105–106
zoogeographic studies, 58–59, 94–95
Zopheridae (zopherid beetle family), 192
distribution patterns, 93
thanatosis, 75, 76
Zopherus
distribution patterns, 92
sanctahelenae, 92
tristus, 76
Zophobas
mealworms, 190
morio, 248–249

ABOUT THE AUTHORS

Arthur V. Evans is a research associate in the Department of Entomology at the National Museum of Natural History, Smithsonian Institution, and the Department of Recent Invertebrates, Virginia Museum of Natural History. He is also the coauthor of *An Inordinate Fondness for Beetles* (California, 2000). James N. Hogue is the manager of biological collections in the Department of Biology at California State University, Northridge, and a research associate at the Natural History Museum of Los Angeles County.

Series Design:	Barbara Jellow
Design Enhancements:	Beth Hansen
Design Development:	Jane Tenenbaum
Illustrator/Cartographer:	Bill Nelson
Composition:	Impressions Book and Journal Services, Inc.
Text:	9.5/12 Minion
Display:	ITC Franklin Gothic Book and Demi
Printer and Binder:	Everbest Printing Company